THE
MODERN
THEME

---

hARPER ⚱ TORCHBOOKS

*A reference-list of Harper Torchbooks, classified
by subjects, is printed at the end of this volume.*

THE MODERN THEME

TRANSLATED FROM THE SPANISH BY JAMES CLEUGH

JOSE ORTEGA Y GASSET

INTRODUCTION BY JOSE FERRATER MORA
HARPER TORCHBOOKS
THE ACADEMY LIBRARY
HARPER & ROW, PUBLISHERS, NEW YORK

Introduction to the Torchbook edition

Copyright © 1961 by Jose Ferrater Mora

Printed in the United States of America

This book was first published in English by C. W. Daniels Co., London, in 1931, and in the United States by W. W. Norton & Company, Inc., in 1933, and it is here reprinted by arrangement with Revista de Occidente of Madrid.

First HARPER TORCHBOOK edition published 1961

CONTENTS

	Introduction to the Torchbook Edition	1
	Preface	9
I	The Concept of the Generation	11
II	The Forecasting of the Future	19
III	Relativism and Rationalism	28
IV	Culture and Life	36
V	The Double Imperative	45
VI	The Two Ironies, or, Socrates and Don Juan	52
VII	Valuations of Life	60
VIII	Vital Values	71
IX	Signs of the Times	78
X	The Doctrine of the Point of View	86

SUPPLEMENTARY

The Sunset of Revolution	99
Epilogue on the Mental Attitude of Disillusion	132
The Historical Significance of the Theory of Einstein	135

v

INTRODUCTION TO THE TORCHBOOK EDITION

by Jose Ferrater Mora

José Ortega y Gasset, more than anyone else, was responsible for launching, in 1917, what during the twenties and early thirties became one of Europe's finest newspapers: *El Sol*, of Madrid. Following Ortega's injunctions, its editors and contributors were intent on turning Spain into a full-fledged member of the European cultural community. One of the consequences of the policy which they adopted was a sharp antipathy toward bull-fighting. The paper did not have a section on bull-fights. Furthermore, its editors staunchly refused advertising which was connected, however remotely, with the *fiesta*. Only when a *matador* or a *picador* was injured, or killed, would the paper print in a corner of one of its last pages an inconspicuous news item under the heading: "The *so-called* national pastime."

Paradoxically enough, Ortega was an *aficionado*. To the dismay of not a few of his friends and colleagues, he was often seen watching the performances of Spanish *toreros*. He became a close friend of some of them. He frankly loathed spinsters who showed more concern for the misery of the bulls than the agonies of the bull-fighters. He impatiently denounced as sheer hypocrites those who campaigned against bull-fighting while lending an eager ear to reports of sadistic crimes. If all this were not enough, he toyed with the project of writing a book, tentatively entitled *Paquiro o de las corridas de toros (Paquiro or About Bull-Fighting)*, which would show the meaning and depth of the "dramatic relationship which has existed for more than two thousand years between the Spaniard and the bull." If we take into account that campaigning against bull-fighting had been in Spain one of the trademarks of the so-called "Europeanizers," it would seem that Ortega, the "Europeanizer" par excellence, was in this respect a far more radical "de-Europeanizer" and "Africanizer" than Unamuno, who had bluntly opposed Spain to Europe, but who apparently had never watched a *corrida* and who never wrote on, and never talked about, bulls and bull-fighters except to express a profound distaste for the whole subject.

I am unable to judge to what extent Ortega was completely sincere, and still less whether he was right, in his defence of the seriousness and profound meaning of bull-fighting. I belong to that section of the Spanish people—more sizeable than most non-Spaniards realize —that has never been concerned with bull-fighting and has never bothered to find out what it is all about. My lack of interest, not to say of competence, in the question is not an effect of partisanship; only of utter indifference. To me and to the vast majority of my Spanish friends bull-fighting is neither good nor bad; it is practically non-existent. But I think that I can understand why Ortega parted company with the writers of his generation on this issue; it is because his intellectual attitude was, in a way, that of a philosophical bull-fighter. He thought of himself as a thoughtful *torero* looking awry and tense at the oncoming bull.

Some of the most characteristic of Plato's metaphors are related to the art of hunting. To pursue a truth, Plato surmised, is like giving chase to a wild beast. The philosopher may be compared to a hunter who follows a trace, a scent, a track. Ortega, on the other hand, was fond of using metaphors drawn from bull-fighting. The best way to tackle a problem, he often contended, is to "catch it by the horn." And problems—real problems, that is—are like bulls: roaring, powerful, and terrifying. Thinkers must be skilful in addition to being thoroughly trained in the rules of the game. They must kill the problems—try to solve them—but also hold them in respect, namely, let the problems be what they are, and not coat them with large doses of optimism or "utopianism." Also, thinkers learn from their predecessors the art of tackling problems in much the way that bull-fighters learn from their seniors the art of fighting bulls. Both thinkers and bull-fighters are indebted to tradition while doing their best to overcome tradition. For when "the hour of truth" strikes, a thinker has to remain alone with his problem, as a bull-fighter stands alone in the ring facing the bull. Nobody can help a thinker at that fateful moment; the circumstances surrounding him are his own entirely. He has to commit himself to his problem in complete solitude, as if he had been thrown into a world where he would be the sole actor, all other men, thinkers or not, being mere spectators waiting for him to emerge as a victor—or to fall as a victim—in a decisive intellectual contest.

Ortega's attitude as an "intellectual *torero*" may be viewed either in a unfavorable or in a favorable light.

Those who take the former stand argue that as soon as a philosopher writes or speaks as if he were alone in the center of a ring, he tends to become overconscious of his intellectual performance.

He is likely to focus more attention upon how to say a thing than upon actually saying it. Ortega, these critics complain, took his audience too much into account. To be sure, he never bowed before it. In fact, he often expressed his contempt for his would-be readers and listeners by claiming that the best of them frequently misunderstood him—if not his words, at least his intentions—and that the worst of them failed to understand him at all. How then can it be maintained that Ortega was excessively influenced by the fleeting reactions of his public? Only by realizing, these critics answer, that a "performer's" low opinion of the qualifications of the public to judge him is a subtle but no less blameworthy way of being haunted by the public. Perhaps the public is not flattered, but then it is tamed—something that is usually done by putting on a show which sacrifices solid and sound performance to mere effect. The proof that such was the case with Ortega, these critics conclude, is obvious in the fact that while he looked at the public so haughtily and contemptuously, in his heart of hearts he was awaiting from it a roaring ovation: a thunderous "*Olé!*"

This unfavorable light on Ortega's intentions and intellectual achievements is at times a consequence of resentment; it often happens that those who cannot do, blame. But in most instances it is a result of a basic misunderstanding of Ortega's aims as a philosopher. At first glance these critics appear to talk sense; there is much in the life and in the work of our author that suggests the image of a brilliant juggler of ideas. But a closer examination of the pattern underlying Ortega's style of thought reveals another, and quite different image.

This underlying pattern is, in fact, a complex tissue of relations between the author's temperament and the intellectual and social milieu in which this temperament developed, between his intellectual training and the cultural conditions in Spain when he started his career as a philosophical writer, and above all between what he wanted to do as an intellectual and what it was possible for him to do with some chances of success. I cannot discuss all these relations in a few pages. But I wish to call attention to a significant fact.

The fact is that it did not take Ortega much time to make a forcible entrance onto the Spanish intellectual scene. From the moment when, just in his early twenties, he began to write and speak publicly, he caught the eye of many of his countrymen. There is no wonder in that: he was saying in Spanish, and in a completely Spanish manner, things that Spaniards were not used to hearing at that time. His thinking was—and always remained—European through and through, but he was not just importing ideas from Europe and clothing them in a Spanish idiom. He was doing something truly astonishing in Spain at the beginning of the twentieth century—producing ideas that, while they were authentically Spanish, sounded also entirely European. Quite naturally, he raised expectations—duly fulfilled in the course of time—that he would become something that Spaniards had long since ceased to dream of: an "exporter" of ideas. Instead of "translating" into Spanish from a foreign tongue, or of confining himself to saying in Spanish things exclusively meant for Spaniards, Ortega was that rare phenomenon in Spain around 1910: someone who could say in Spanish things that non-Spaniards might hail as important or challenging.

As a consequence, Ortega found himself placed in "the center of the ring." Friends and enemies alike grew into the habit of waiting for what "the philosopher" would say or write, and there was obvious disappointment when he refused to air his views. If Ortega had been only a philosopher he might not have needed to worry about causing any such disappointments. He might have confined himself to fulfilling the task that Santayana assigned to philosophers: to "watch stars that move slowly." But Ortega was far from being "only a philosopher." At any rate, he was not a philosopher in the conventional sense of this word, namely, an absent-minded academic who might *also* be a citizen of his country, a lover, a concert-goer, or a pulp magazine reader. For him, being a philosopher included being a Spaniard, a political thinker, a writer, and, of course, a modern man. Furthermore, even if Ortega had wished to be a philosopher independently of anything else, he could not have limited himself to abstruse philosophizing. It was all right for Edmund Husserl to do practically nothing but philosophy in Göttingen or in Freiburg. But it was impossible—or rather, as Ortega put it, "indecent"—to act as a purely technical philosopher in Madrid in 1910, 1923, or even 1931. Ortega was expected to say something, "to say his word," on topics which philosophers *as philosophers* usually evade or treat only perfunctorily. He was expected to say something about love, women, Spain, the poetry of Anne de Noailles, the Roman Empire, the "population explosion," the theory of relativity, and what not. He was expected above all to say his word on political and social issues in contemporary Spain. Although uncommitted (or perhaps because uncommitted) to any political group, he was seen as an incomparable political "enlightener." He was likely to treat such issues more deeply than any of the professional political commentators, and more specifically than any of the traditional philosophers of history. And since political and social issues in one country can be fruitfully explored only when they are placed within the context of the historical situation of the world at large, Ortega was expected to say something—indeed, a great deal—about this larger situation and the consequences thereof.

If Ortega had only been "expected to say" his word on so many issues, there would be no reason why he should not have decided to keep silent on most of them. But he was expected to say his word only because his readers and his listeners understood the man even before they had understood, or for that matter heard of, his philosophy. They saw in Ortega a thinker who had a definite vocation (what he later described as "the vocation for clarity"). I do not know whether they also saw in him one who would eventually assert that men live authentically only in so far as they loyally accept their circumstances, but they treated him as if he were going to assert something like this. At any rate, it was a happy coincidence that Ortega was expected "to say his word," because saying his word, and saying it clearly was indeed his real vocation. And this activity had to be performed in view of the circumstances surrounding him and his fellow-citizens. Ortega recognized this fact and made it a milestone of his philosophy. As early as 1914 he wrote that men must live in awareness of their circumstances and must struggle to "save" these circumstances; which, from a philosopher's point of view, means: "give an account of them." Now, Ortega's own circumstances can be summarized as follows: he was a Spaniard earnestly concerned with the past, the present, and the future of his country; he was a man who felt at home in Europe (particularly in Germany); he was a modern man deeply involved in the problems of contemporary civilization; he was endowed with an unusual power for thinking clearly and writing brilliantly. He could not do away with any of these "circumstances"; rather, he felt that he had to make the most of them. He believed that it was his task to clarify the problems that these circumstances raised and certainly not as an innocent pastime in which one may occasionally indulge, but as a full-time job. If we are ready then to understand Ortega's intentions, and the reasons supporting them, we will no longer deplore the fact that his philosophy was not expressed like Husserl's or Rudolf Carnap's. We will not consider it too alarming that Ortega had a brilliant style. For one thing, a brilliant style is a most refreshing thing nowadays, when some philosophers have come to believe that it is their duty to be as dull as ditch-water. We will understand, in short, that acting as if he were an "intellectual *torero*" was for Ortega less a display of arrogance than a consequence of a thoughtful purpose conditioned by the cultural circumstances of his life.

It may still be claimed that Ortega yielded too readily to the intellectual incitements of the moment, and that it would have been wiser for him to stop talking about love and politics (not to say bull-fighting) and devote all of his time to "straight philosophical investigation." Yet, three facts must be considered before making hasty pronouncements on this issue. First, that the many themes which Ortega touched upon he always discussed in an undeniably philosophical vein. Second, that there has been far more "straight philosophical investigation" in Ortega than most of his readers have suspected. Third, and particularly important here, that although he treated a great variety of subjects, one of these crops out constantly in his works and never drops out completely: the subject "man and history."

Ortega's obvious preference for this subject is grounded on his idea of man as a reality fundamentally historical. But for Ortega history, as well as human life, is not a shapeless succession of events: it is a "system." Ortega's analyses of history are, therefore, philosophical through and through. They are so to such a degree that some readers feared that he might yield to the common temptation among philosophers of considering history at large as a superstructure of the history of philosophy. Fortunately, such fears proved groundless. If Ortega's strong sense for historical realities—so forcefully expressed in the essay, herewith included, "The Sunset of Revolution"—had not been enough to prevent him from becoming a mere "historian of ideas" under the guise of a philosophical historian, his metaphysical convictions would have come to the rescue. For Ortega emphasized the historical character of philosophy itself. Philosophy is, he thought, born at a certain date, within certain historical circumstances, and may eventually vanish without history necessarily coming to an end. This does not mean that he considered philosophy as history. To assert that philosophy is a historical event is not yet to say that philosophical truth is relative to historical conditions. The relative and the absolute interact and play a fascinating and subtle "dialectical game."

A confirmation of Ortega's equal opposition to the relativism of pure "historicists" and to the absolutism of pure "rationalists'" can be found in *The Modern Theme*. Ortega here expressed certain of his philosophical views with exemplary lucidity. Doctrines such as that of the idea of generation and its function in human history, of the type of relations to be established between cultural values and vital spontaneity, and of the metaphysics and theory of knowledge that he often called "perspectivism" can be found in not a few of Ortega's books and essays, where they are sometimes developed in a more convincing manner than in *The Modern Theme*. But only in this book are these conceptions woven together to serve as guide-posts for an understanding of the whole modern period. *The Modern Theme* is in many senses a book typically Ortegean: swarming with insights into the condition of man in his historical environment, of interpretations of the past, and anticipations of the future. Enthusiasm is tempered here by disillusion, and in this respect the book reflects not only a substantial portion of Ortega's intellectual experience, but also the conditions of the times when it was composed. It is not one of Ortega's most profound philosophical works—as it happens, he wrote his most permanent contributions to philosophy during the last ten years of his life, and only now are they being published as a series of posthumous works. But *The Modern Theme* contains many seeds that developed later and that found their proper place within a carefullly worked out "metaphysics of human life" (the expression that best summarizes Ortega's pervasive philosophical theme).

Many philosophers are not reliable expositors of their own thought; critics and teachers must mediate between them and those who try to understand what the philosopher intended to say. Some philosophers, on the other hand, are their own best expositors. This must not be construed as a reflection on the merits or defects of philosophers; both groups include major philosophers as well as minor ones. Ortega was a philosopher who could present his thoughts better than anyone else. It is still possible, not to say desirable, to talk about Ortega's philosophy: summarize, scrutinize, and criticize it. But it is difficult to present it to a reader, and in particular to the "cultivated general reader," in a better garb than the one the author himself gave it. At any rate, it would be a waste of time to offer here an exposition of *The Modern Theme;* the reader may rest assured that he has at hand the best one available. But I wish to caution against considering the present book as a complete and sufficient account of Ortega's philosophical thought. The true meaning of some of the crucial concepts developed in *The Modern Theme* can be fathomed only by fitting them together with others handled by the author, and also with the very same concepts as he worked them out later. On various occasions Ortega emphasized a certain doctrine in order to reveal the vacuity of the doctrine opposing it. In *The Modern Theme,* for example, there is a spirited defence of "vital values" that may induce some candid readers to conclude that Ortega entirely subscribed not only to "vitalism," but even to some sort of crude "biologism." This proves not to be so as soon as we take into account Ortega's further qualifications of his "vitalistic" doctrine. Other examples might be cited, but the one above will, I hope, suffice to forestall hasty conclusions as to Ortega's "definitive" philosophical orientations. If these precautions are kept in mind, the present book may rightly be regarded as a very adequate introduction to Ortega's ideas and style of thinking. At any rate, it is characteristic of Ortega's in being a fascinating and delightfully readable piece of prose.

Bryn Mawr College
Bryn Mawr, Pennsylvania
January, 1961

PREFACE

THE first part of this book contains a somewhat amplified draft of the lecture with which I began my ordinary university course for the year 1921-22.

In giving the lecture its present form I took advantage of the detailed and extremely accurate notes taken in the hall itself by a member of my audience, my esteemed friend Don Fernando Vela.

This discourse is now to be submitted to a less specialised public than that which was assembled at the university, and accordingly I thought it essential to be rather more explicit in regard to certain points, which might be less easily assimilable by readers unaccustomed to philosophic study. Such was the only reason for the amplification I have made of the original text.

Certain appendices follow, and deal with less general questions. These questions are, however, all associated with the main theory explained in the lecture itself. My own interest is centred particularly in the appendix, which gives a brief account of a philosophic interpretation of the general significance latent in Einstein's theory of physics. I believe that for the first time attention is there drawn to a definitely ideological quality inherent in that theory which contradicts the interpretations hitherto current.

CHAPTER I

THE CONCEPT OF THE GENERATION

THE most important point about a scientific system is that it should be true. But the explanation of a scientific system involves a further postulate: besides being true it must be understood. I am not for the moment referring to the difficulties imposed upon the mind by a scheme of abstract thought, especially if unprecedented, but to the comprehension of its fundamental tendency, of its ideological significance, I might almost say, of its physiognomy.

The scheme of thought with which we are now to be concerned claims to be true, in other words to reflect the character of phenomena with fidelity. But it would be utopian and for that reason false to suppose that in order to make good its claim our system is to be directed exclusively by phenomena and give its undivided attention to its mere context. If the philosopher concentrated simply upon material objects philosophy would never be anything but primitive. Superimposed, however, upon material phenomena, the investigator finds the thoughts of other people, the whole body of human meditation in the past, innumerable traces of previous explorations, the signs of journeys attempted through the eternal jungle of problems, still virgin in spite of repeated violation.

Every philosophical enquiry has therefore to take two data into consideration: the nature of phenomena and the speculations to which phenomena have given rise. Such a collaboration with the theories of the past ensures, at any rate, freedom from errors already committed, and gives a progressive character to the succession of philosophical systems.

Now, the thought of any age can assume two opposite attitudes to what has been thought in other ages. Especially is this the case in regard to the immediate past, which is always the most powerful influence and contains in concentrated form everything anterior to the present. There are in fact some ages in which thought regards itself as growing out of seeds already sown, and others which are conscious of the immediate past as of something in urgent need of radical reform. The first-named are the ages of pacific, the second those of militant philosophy, the aspiration of which is to destroy and completely supersede the past. Our own age is of the latter type, if we understand by "our own age" not that which has just come to an end but that which is just beginning.

When thought found itself compelled to adopt a bellicose attitude to the immediate past the intellectual world was divided into two camps. On the one side stood the great majority, which clung to established ideology; on the other a small minority of spiritual scouts, vigilant souls, who had a glimmering of distant tracts of territory still to be invaded. This minority is doomed never to be properly understood; the gestures which the vision of the new dominions calls forth from it cannot be rightly interpreted by the main body advancing behind and not yet in possession of the height from which the "terra incognita" is being examined. Hence the minority in the van is in continual danger, both from the new districts it has to conquer and from the rank and file harassing its rear. While it is constructing the new it has to defend itself from the old, and wield at one and the same time, like the renovators of Jerusalem, both the spade and the sword.

This division is deeper and more absolute than is generally believed. I will try to show in what sense it subsists.

We attempt, by means of history, to understand the changes which take place in the minds of men. For this purpose we have first to observe that these changes are not all of the same kind. Certain historical phenomena depend on others, more fundamental, and themselves independent of the former. The idea that everything has a bearing on everything else, that everything depends on everything else, is a loose exaggeration of the mystics and ought to be repugnant to anyone determined to see his way clearly. On the contrary, the body of historical reality exhibits a complete hierarchy in its anatomy, an orderly succession of subordinate parts and an equally successive interdependence between the various classes of facts. Accordingly, changes of an industrial or political nature are superficial: they depend upon ideas, upon contemporary fashions in morals or aesthetics. But ideology, taste and morality in their turn are no more than consequences or demonstrations of the root feeling that arises in the presence of life, the sensations of existence in its undifferentiated totality. What we are going to call vital sensibility is the primary phenomenon in history and the first we should have to define in order to understand a particular age. Yet, when a variation in sensibility appears only in a single individual, it has no true historical significance. It has been customary for the field of the philosophy of history to be disputed by two tendencies which are both, in my judgment, and if I may say so without at the moment having any intention of going into the question in detail, equally erroneous. There has been both a collectivist and also an individualist interpretation of historical reality. For the former the process of history is in substance the work of widely diffused masses of mankind; for the latter the makers of history are exclusively individuals. The active, creative nature of personality is, in fact, too evident for the collectivist picture of history to be acceptable. Mankind in the gross is merely receptive: it is content simply to show favour or resistance to men of marked and enterprising vitality. On the other hand, however, the isolated individual is an abstraction. Historical life is social life. The life of the outstanding individual consists precisely in a comprehensive influence on the generality. "Heroes," then, cannot be separated from the rest of the world. There emerges here a duality which is essential to the process of history. Humanity, in all the stages of its evolution, has always been a functioning organism, in which the more energetic members, whatever form their energy may take, have operated upon the remainder and given them a distinct configuration. This circumstance implies a certain basic sympathy between the best type of individual and the generality of mankind. An individual whose nature differed completely from that of the mass would not produce any effect whatever upon it: his actions would pass over the surface of the society of his age without arousing the least response from it and therefore without entering the general process of history. To one extent and another such occurrences have been fairly frequent; and history has to note in the margin of its main text the biographical details of "extravagant" characters of this kind. Like all other biological sciences history has a department devoted to monstrosities, a teratology, in fact.

The changes in vital sensibility which are decisive in history, appear under the form of the generation. A generation is not a handful of outstanding men, nor simply a mass of men; it resembles a new integration of the social body, with its select minority and its gross multitude, launched upon the orbit of existence with a pre-established vital trajectory. The generation is a dynamic compromise between mass and individual, and is the most important conception in history. It is, so to speak, the pivot responsible for the movements of historical evolution.

A generation is a variety of the human race in the strict sense given to that term by naturalists. Its members come into the world endowed with certain typical characteristics which lend them a common physiognomy, distinguishing them from the previous generation. Beneath this general sign of identity, individuals of so diverse a temper can exist that, being compelled to live in close contact with one another, inasmuch as they are contemporaries, they often find themselves mutually antipathetic. But under the most violent opposition of "pros" and "antis" it is easy to perceive a real union of interests. Both parties consist of men of their own time; and great as their differences may be their mutual resemblances are still greater. The reactionary and the revolutionary of the nineteenth century are much nearer to one another than either is to any man of our own age. The fact is, that whether they are black or white the men of that generation belong to one species, while in our own persons, whether we are black or white, are the beginnings of a further and distinct variety of mankind.

The disputes between "pros" and "antis" are less important in the orbit of a generation than the constant interval separating the elect and the masses. When one is confronted by current theories which ignore or deny this evident difference in historical value between the two classes one feels justifiably tempted to exaggerate it. However, the assumption underlying those very discrepancies in stature is that individuals arise from an identical point of departure, a common plane above which some achieve greater heights than others, and which plays the part of the sea-level in topography. And each generation does in fact represent a certain vital altitude at which existence is conscious, in a certain sense, of being determined. If we consider the evolution of a race in its entirety we find that each of its generations appears as a moment in its vital process or as a pulse-beat in its organic energy. And each pulse-beat has a peculiar, even a unique, physiognomy; it is an inconvertible palpitation in the series of pulse-beats, like each note in the development of a melody. We may equally well picture each generation by means of the image of a biological projectile launched into space at a definite moment and with pre-determined force and direction. In such determination all its elements, both the most valuable and the most normal, participate.

Nevertheless, it is clear that we are at present only designing forms or colouring illustrations which serve our purpose of isolating the really significant fact through which the concept of the generation proves its truth. This is simply that the generations are born one of another in such a way that the new generation is immediately faced with the forms which the previous generation gave to existence. Life, then, for each generation, is a task in two dimensions, one of which consists in the reception, through the agency of the previous generation, of what has had life already, e.g., ideas, values, institutions and so on, while the other is the liberation of the creative genius inherent in the generation concerned. The attitude of the generation cannot be the same towards its own active agency as towards what it has received from without. What has been done by others, that is, executed and perfected in the sense of being completed, reaches us with a peculiar unction attached to it: it seems consecrated, and in view of the fact that we have not ourselves assisted in its construction, we tend to believe that it is the work of no one in particular, even that it is reality itself. There is a moment at which the concepts of our teachers do not appear to us to be the opinions of particular men, but truth itself come to dwell anonymously upon the earth. On the other hand our spontaneous sensibility, the thoughts and feelings which are our private possessions, never seem to us properly finished, complete and fixed, like a definite object: we regard them more as a species of internal flux, composed of less stable elements. This disadvantage is compensated by the greater expansiveness and adaptability to our own nature always characteristic of spontaneity.

The spirit of every generation depends upon the equation established between these two ingredients and on the attitude which the majority of the individuals concerned adopts towards each. Will that majority surrender to its inheritance, ignoring the internal promptings of spontaneity? Or will it obey the latter and defy the authority of the past? There have been generations which felt that there was a perfect similarity between their inheritance and their own private possessions. The consequence, then, is that ages of accumulation arise. Other times have felt a profound dissimilarity between the two factors, and then there ensue ages of elimination and dispute, generations in conflict. In the former case the young men coming to the front coalesce with the old and submit to them: in politics, in science and in the arts the ancient *régime* continues. Such periods belong to the old. In the latter case, since there is no attempt at preservation and accumulation, but on the contrary a movement towards relegation and substitution, the old are swept away by youth. Such periods belong to the young and are years of innovation and creative struggle.

This rhythm of ages of senescence and ages of rejuvenation is a phenomenon so patent in a long view of history that it is surprising not to find it recognised by everyone. The reason for such lack of recognition lies in the fact that there has never been any attempt to found, even formally, a new scientific system which might be called "Metahistory" and which would bear the same relation to concrete histories as physiology to the clinic. One of the most interesting of "metahistorical" investigations would consist in the discovery of the great historical rhythms. For there are others no less evident and fundamental than that already referred to, for example, the rhythm of sex. There is in fact a pendulum movement latent in history which swings from ages subjected to the dominant influence of respectability to ages that surrender to the yoke of the female principle. Many institutions, customs, ideas and myths, hitherto unexplained, are illuminated in an astonishing manner when the fact is taken into account that certain ages have been ruled and modelled by the supremacy of women. The present moment, however, is not a fitting opportunity to enter upon that particular question.

CHAPTER II

THE FORECASTING OF THE FUTURE

IF the essence of each generation is a particular type of sensibility, an organic capacity for certain deeply-rooted directions of thought, this means that each generation has its special vocation, its historical mission. It is under the strictest compulsion to develop those tiny seeds and to give the existence of its environment a form corresponding to the pattern of its own spontaneity. But generations, like individuals, sometimes fail in their vocation and leave their mission unachieved. There are in fact generations which are disloyal to themselves and defraud the cosmic intention deposited in their keeping. Instead of resolutely undertaking their appointed task they remain deaf to the urgent summons of the vocation that is really theirs and prefer a supine reliance on ideas, institutions and pleasures created by their forbears and lacking affinity with their own natures. It is obvious that such a dereliction of historical duty cannot go unpunished. The guilty generation drags out its existence in perpetual division against itself, its essential life shattered.

I believe that in the whole of Europe, but more particularly in Spain, the present generation is one of this derelict type. Seldom have men lived in greater mental confusion, and perhaps never before has humanity endured an unsuitable setting with such docility; a setting which is a survival from the past and does not tally with the internal rhythm upon which it is imposed.

Hence arises the characteristically modern apathy in matters, for example, of politics and art. Our institutions, like our theatres, are anachronisms. We have neither had the courage to break resolutely with such devitalised accretions of the past nor can we in any way adjust ourselves to them.

In these circumstances a system of thought, such as I have for some years been expounding from this Chair, is not easily to be understood in its ideological intention, in its internal physiognomy. It aims, perhaps unsuccessfully, at fulfilling as conscientiously as possible the historical imperative of our generation. But our generation seems profoundly determined to misconceive the suggestions of our common destiny. I am compelled to conclude that even the most gifted among us, with very rare exceptions, have not the slightest suspicion that the pointer in the compass of modern Western sensibility is veering through at least ninety degrees. I have therefore considered it necessary to anticipate, in this first lecture of mine, some part of what constitutes, in my judgment, the essence of the theme with which modernity has to deal.

How has such an utter lack of recognition been possible? When, during a conversation on politics with some "advanced," "radical" or "progressive" contemporary —let us put as favourable a case as we can—the inevitable disagreement comes to light, our opponent thinks that this disagreement on questions of administration and State affairs is properly to be called a divergence of political views. But he is wrong: our political difference is a very secondary matter and would be completely unimportant if it did not serve the purpose of summarily indicating the existence of a far deeper dissension. Our divisions in politics are not so considerable as those prevailing in our guiding principles of thought and feeling. We are separated, long before we come to constitutional doctrine, by diverse systems of biology, physics, historical philosophy, ethics and logic. The political position of any one of our contemporaries is determined by certain ideas which we have all received from those who were once our masters. They are ideas which reached their full maturity about the year 1890. Why is it that people are content to rely upon inherited systems of thought, in spite of having to observe how repeatedly they conflict with the spontaneity of the age? Men prefer service, without real allegiance, under outworn banners, to compliance with the painful effort of revising inherited principles and setting them in accord with their own deepest feelings. It does not matter whether people are Liberals or Conservatives: in either case they are stragglers from duty. The destiny of our generation is not to be liberal or to be conservative, but merely to put that ancient dilemma out of mind altogether.

People who are under the obligation, by reason of their eminent intellectual qualities, of assuming responsibility for the conduct of our age, have no excuse for living, like the masses, on a derivative level, harnessed to the superficial caprices of every moment, without attempting to find some disciplined and comprehensive orientation towards the courses of history. For history is not a mere series of accidents beyond the control of forecast. It is not of course possible to foretell the various events that to-morrow will bring forth; neither, indeed, would such a prediction be of any real interest. But on the other hand it is perfectly possible to foresee the characteristic thought of the immediate future, to anticipate the general features of the period that will succeed one's own. In other words, a thousand unpredictable accidents occur in any given period; but the period itself is not an accident. It possesses a fixed and unmistakable structure. Its case is similar to that of individual destinies: no one knows what is going to happen to him to-morrow; but he does know his own character, his own desires and his own powers, and hence the way in which he will react to whatever accidents may befall him. Every life has a pre-established normal orbit, in the course of which accident, without essentially deflecting the orbit in question, traces certain sinuosities and indentations.

History has room for prophecy. And more than this: the labour of history is only scientific in proportion to the place that prophecy can occupy in it. When Schlegel alleged that the historian was an inverted prophet he expressed an idea that is as profound as it is true.

The view of life held by antiquity makes history, strictly speaking, worthless. For the ancients, existence was determined by fate. Historical events were externally motived contingencies which affected successively this or that individual or nation. The production of a work of genius, financial crises, political changes and wars were phenomena of an identical type, which can be symbolised by the tile falling on the head of a passer-by. On this interpretation historical process is a series of haphazard shocks without rhyme or reason. Historical science is therefore an impossibility, science only being possible where some ascertainable law exists, a formula which, in virtue of the sense that it implies, can be understood.

Life, however, is not an externally-motived process in which there is nothing but a mere aggregation of contingencies. Life is a series of events ruled by law. When we sow the seeds of a tree we foresee the whole normal course of its existence. We cannot foresee, indeed, whether lightning will come and cut it down with the scimitar of fire that hangs by the side of some cloud; but we know that the seed of the cherry will not bear the foliage of the poplar. In the same way the Roman people is a collection of vital tendencies which go on gradually developing in time. In each stage of this development the next phase is already implicit. Human life is an internally motived process in which the essential events do not occur as directed from outside upon the subject of experience, an individual or a nation, but evolve from within it, as fruits and flowers evolve from seeds. It is in fact an accident that in the first century before Christ there lived a man endowed with the singular genius of Cæsar. But what Cæsar, with his singular genius did brilliantly, ten or twelve other men, whose names we know, would also have done, no doubt less brilliantly and less thoroughly. A Roman of the second century before Christ could not foresee the individual destiny which was the life of Cæsar; but he could have prophesied that the first century before Christ would be a "Cæsarean" age. By one name or another "Cæsarism" was a generic form of public life which had been maturing since the time of the Gracchi. Cato prophesied very clearly the destinies carried by the future that was immediately to succeed his own age. Human existence being life in the most literal sense, that is, a process internally motived and subject to the operation of a law of development, it is possible to admit a science of history. In the last analysis science is nothing but the effort we make to understand anything. And we have understood a situation historically when we perceive that it arises necessarily from another situation anterior to it. What kind of necessity is here meant—physical, mathematical or logical? None of these: the necessity in question is related to such classifications, yet it has a character peculiar to itself:

it is psychological necessity. Human life is in an eminent degree psychological life. When we are told that Pedro, who is an honest fellow, has killed his neighbour, and we next discover that his neighbour has dishonoured Pedro's daughter, we have sufficiently well comprehended the nature of his homicidal act. Our comprehension is due to our recognition that one act proceeds from the other, vengeance from dishonour, in an unmistakable trajectory and on evidence of the same kind as that guaranteed by mathematical truth. But on the same evidence, viz., the knowledge of the daughter's dishonour, we could predict, before the actual crime, that Pedro would kill his neighbour. In this case it is perfectly clear that to prophesy the future we make use of the same intellectual operation as serves us to understand the past. In both directions, backwards or forwards, we are bound to admit the existence of a single manifest psychological curve, just as, when we find the segment of an arc we are able to complete the entire figure without hesitation. I believe, then, that the expression I used a little while ago to the effect that historical science is only possible in the proportion that prophecy is possible, will not now appear a rash one. When historical thought matures the capacity to forecast augments with it.*

But leaving aside all the secondary questions that the conscientious exposition of this theory may suggest, let us confine ourselves to the possibility of foreseeing the immediate future. How are we to proceed with our design?

It is evident that the near future will be born of ourselves and will consist in the extension of what is essential

* My readers will observe that this doctrine of a possible anticipation of the future bears hardly any relation to that of "Historical Prophecy" recently proclaimed by Oswald Spengler.

and not contingent in us, normal and not the result of chance. Strictly speaking, then, it would be enough if we descended into our own hearts and eliminating all individual projects, private predilections, prejudices and desires prolonged the directions of our appetites and essential tendencies to the point at which we could perceive that they coalesced in a single type of life. But I am well aware that this operation, simple as it looks, is not so in the case of persons unaccustomed to the rigours and precisions of psychological analysis. There is nothing less customary, in fact, than such a forcing of the mind back upon and into itself. Man has been formed in his struggle with external nature and it is only easy for him to discern phenomena outside himself. When he looks within vision is clouded and he grows dizzy.

I believe, however, that there is a further objective procedure which will make it possible to discover symptoms of the future in the present.

I said above that the body of the ages of mankind possesses an anatomy subsisting on hierarchic principles, and that there are in it certain primary activities and further secondary ones derived from them. Accordingly, such characters as have within a period of twenty years suceeded in manifesting themselves in the secondary activities of life, which are the most patently perceptible, will have already commenced to affect the primary activities. Politics, for example, is one of the more secondary functions of historical life, in the sense that it is a mere consequence of all the rest.* When a certain spiritual state succeeds in colouring political movements it has already passed through all the other functions of the historical organism. Politics result from the gravita-

* On this point historical materialism, though its motives seem to me inadmissible, is in the right.

tional interaction of masses. Now, in order that a modification of historical consciousness may reach the mass, it must have previously influenced the choice minority. The members of the latter are of two classes: men of action and men of contemplative nature. There is no doubt that the new tendencies, not yet at their full strength, will be perceived by the contemplative natures earlier than by the active. The pre-occupation of the moment prevents the man of action from feeling the first vague stirrings of the breeze that is not yet ready to fill the sails of his practical temperament.

It is in the realm of pure thought, therefore, that the earliest faint signs of the coming age can be traced. They are the light ripples caused by the first few puffs of wind on the calm surface of the pool. Thought is man's most fluid possession; and accordingly it yields freely to the slightest variations in his vital sensibility.

To sum up: the science of to-day is the magic vessel into which we have to look to obtain a glimmering of the future. The modifications, which may appear to be only technical, that modern biology, physics, sociology or prehistory are producing, through their experimental work, in the whole fabric of philosophy, are the preliminary gestures of the new age. The extremely delicate subject matter of science is sensitive to the least vacillations of human vitality, and is capable of acting at the present moment as a register, on a very small scale, of phenomena which will, with the passage of years, loom gigantic upon the stage of public life. Anticipation of the future can, then, rely upon a precisely adjusted instrument, resembling the seismograph, which indicates by a slight tremor the occurrence of a catastrophic earthquake thousands of miles away.

Unless our generation desires to lag behind its own destiny it must adopt some kind of orientation towards the general character of modern science instead of clinging to the political philosophy of the moment, which is entirely anachronistic, and a mere echo of the voice of a dead sensibility. On what men are beginning to think to-day depends how they will live in the market-places to-morrow.

Fichte, for the benefit of his own time, attempted a similar task to that of these lectures in the celebrated series of discourses afterwards published in a volume entitled "The Characteristics of This Age." I propose to restrict the terms of my contract to a summary description, which I shall now attempt, of my idea of the principal theme of our own.

CHAPTER III

RELATIVISM AND RATIONALISM

I HAVE assumed throughout my discourse the existence of an intimate affinity between scientific systems and generations or ages. Does this mean that science, and particularly philosophy, is a body of opinions that only hold good for a certain period? If we accept the transitory nature of all truth in this way we shall find ourselves enrolled in the band of professors of the relativist theory, which is one of the most typical productions of the nineteenth century. We talk of escaping from this age, but we are merely relapsing into it.

This question of truth, which appears to be incidental and purely technical in character will take us directly to the very root of the modern theme.

The term, truth, conceals a highly dramatic problem. Truth, if it is to give an adequate reflection of the nature of phenomena, must be complete in itself and invariable. But human life in its multiform development, that is to say, in history, has constantly changed its mind, consecrating as true whatever it happened to be in favour of at the moment. How are we to reconcile these two opposing data? How can we admit truth, which is complete in itself and invariable, to the society of human vitality, which is essentially mutable and varies from individual to individual, from race to race, and from period to period? If we wish to keep to living history and pursue its suggestive undulations we must renounce the supposition that truth is ascertainable by man. Every individual has his own opinions, more or less permanent, which are "for him" the truth. Upon them he founds his domestic fireside, which keeps him warm on the surface of existence. "The" truth, then, does not exist: there are only truths "relative" to the frame of mind of the person considering the matter. Such is the relativist theory.

But this renunciation of truth, so lightly undertaken by the relativist theory, is a more difficult business than may at first appear. It is claimed that by this means a lofty impartiality of outlook is obtained upon the multiplicity of historical phenomena; but what is the price? In the first place, if truth does not exist, relativism cannot take itself seriously. Secondly, belief in truth is a deeply-rooted foundation of human life; if we remove it, life is converted into an illusion and an absurdity. The operation of removal is itself devoid of common sense and philosophic value. Relativism is, in the long run, scepticism, and scepticism, when its justification is that it opposes all speculative theory, is in itself a theory of suicidal character.

The relativist tendency is no doubt inspired by a praiseworthy attempt to respect the glorious independence natural to all vitality. But it is an attempt which fails. As Herbart said, every sound originator is a sceptic, but no sceptic is ever anything but an originator.

A deeper current in European consciousness since the Renaissance is the opposite tendency: rationalism. Proceeding by an inverse method, rationalism, for the sake of retaining truth, renounces life. Both tendencies meet in the positions attributed by the popular couplet to the two Popes, seventh and ninth of the name:

>   Pius abandons faith to keep the See.
>   Pius abandons See to keep the faith.

Truth being one, absolute and invariable, cannot be attributed to our own individual personalities, which are corruptible and mutable. We must assume, beyond the differences which exist among men, a sort of abstract type, common to the European and the Chinese, to the contemporary of Pericles and to the "grand seigneur" of the age of Louis XIV. Descartes called this common basis of mankind, which is exempt from individual variations and peculiarities, "reason," and it was called by Kant "the rational entity."

The schism thus effected in man's personality should be carefully noted. On the one side stands everything vital and concrete in his being, his breathing and historical reality. On the other, that rational nucleus which enables us to attain truth, but which nevertheless has no life. It is an unreal phantom, gliding immutably through time, alien to the vicissitudes which are a symptom of vitality.

Yet it is not clear why, in these circumstances, reason has not discovered the universe of truths. How is it the process is so slow? How comes humanity to yield a thousand times to the bemusing embraces of the most diverse errors? How are we to explain the multiplicity of opinions and tastes which have dominated history as ages, races and individuals succeed one another? From the point of view of rationalism history, with its incessant shocks to anticipation, lacks common sense and is properly the history of the obstructions which block the path of the emergence of reason. Rationalism is anti-historical. In the system of Descartes, the father of modern rationalism, history finds no place, or rather, is relegated to the dock. "Everything that reason conceives," he says in the Fourth Meditation, "It conceives rightly, and there is no possibility of error. Whence, then, come my errors? They come simply from the fact that the will being so much greater in volume and extent than the understanding I do not confine the former to the limits of the latter but extend it to things that I do not understand; having no preferences of its own in this field it very easily goes astray and selects the false as the true and the bad as the good: such is the cause of my mistakes and offences."

It follows that error is an offence of the will, not an accident; perhaps it may even be the inevitable fate of intelligence. If it were not for the offences of the will the first man would long ago have discovered all the truths that are accessible to him; in the same way there would have been no variations in opinion, law and custom; in short, there would have been no history. But as there has been such a thing as history, we have no resource but to attribute it to offence. History would then be substantially the history of human error. There can be no attitude more anti-historical, none more anti-vital than this. History and life are burdened with a negative significance and given a strong savour of crime.

The case of Descartes is an exceptionally pointed illustration of the foregoing observations on the possibility of forecasting the future. His contemporaries, too, saw nothing in his work, at first, but an innovation of purely scientific interest. Descartes was proposing to substitute certain physical and philosophic doctrines for others and the attention of his audience was directed solely to the question whether the new doctrines were acceptable or erroneous. The same situation arises to-day in the case of the theories of Einstein. But if, abandoning for the moment our attention to the question of value and suspending our judgment as to the truth or falsity of the Cartesian system, we had considered it simply as an initial symptom of a new sensibility, as a budding manifestation of a new epoch, it would have been possible to discover in it the silhouette of the future.

For what was the irreducible core of the physical and philosophic thought of Descartes? It was a declaration of the dubiety and consequent negligibility of every idea or belief not constructed by "pure intellection." Pure intellection, or reason, is nothing else but our understanding functioning in the void, without let or hindrance, in contact with itself, and controlled only by its own internal standards. For example, to take vision and imagery, a point is the smallest spot that we can effectually perceive. For pure intellection, however, a point is merely what is fundamentally and absolutely smallest: it is something infinitely small. Pure intellection, the *"raison"* of Descartes, can only operate among superlatives and absolutes. When it sets itself to think of a point it cannot stop at any dimension short of the ultimate. This is the geometric mode of thought, the *"mos geometricus"* of Spinoza, the "pure reason" of Kant.

The enthusiasm of Descartes for the constructions of reason led him to effect a complete inversion of the perspective natural to mankind. The immediate and evident world we observe with our eyes, touch with our hands, listen to with our ears, is composed of qualities: colours, resistances, sounds and so on. This is the world in which man has always lived and always will live. But reason is not capable of dealing with qualities. A colour cannot be thought or defined. It must be seen, and if we wish to speak of it we must make contact with it. In other words, colour is irrational. On the contrary, number, even that species of number called by mathematicians "irrational," is a constituent element of reason. Without doing more than making contact with itself it can create the universe of quantities by the use of definite and clearly-outlined ideas.

With heroic audacity Descartes decides that the true world is the quantitative, the geometrical: the other, the qualitative and immediate world that surrounds us in all the plenitude of its beauty and suggestive force, is dismissed, and assumed to be, in a way, illusory. It is true that the illusion is so solidly founded upon our own nature that we cannot avoid it by knowing it for illusion. The world of colour and sound continues to seem as real as before we discovered its imposture.

The Cartesian paradox is the foundation of modern physics. We have been brought up on it and it costs us an effort, now, to see how utterly unnatural it is and to put back our boundary stones where Descartes found them. It is clear, however, that so complete an inversion of our spontaneous perspective was not, for Descartes and the generations which followed him, an unforeseen result suddenly arrived at by proofs that did not admit of doubt. On the contrary the process began with a more or less vague desire that the nature of phenomena should be of a certain type, and it continued in a search for proofs to demonstrate that their nature should be, in fact, such as was desired. I must not be interpreted as in any way contending that the proofs found are illusory: I merely wish to point out that it is not the proofs that seek out and attach themselves to us, but we who go in search of them, in the prosecution of our previous designs. No one supposes that Einstein was surprised to discover, one fine day, that he had to recognise the existence of four dimensions in the world. For thirty years many men of keen intelligence had been postulating a four dimensional physics. Einstein deliberately went in search of it, and as the desire to find it was not incapable of fulfilment he found what he was looking for.

The physics and philosophy of Descartes were the

first manifestations of a new spiritual state which, a century later, came to overspread all the forms of human life and predominated in the drawing-room, the law court and the market-place. The convergence of the features of this spiritual state produced the sensibility which is specifically "modern." Mistrust and contempt of everything spontaneous and immediate. Enthusiasm for all the constructions of reason. To the Cartesian or "modern" man the past will be antipathetic because then things were not done "*more geometrico*." Accordingly he will consider traditional political institutions stupid and unjust. As opposed to them, he believes he has discovered a definitive social order arrived at deductively by means of reason. It is a schematically perfect constitution in which it is assumed that men are rational entities and nothing else. This assumption being granted— "pure reason" has always to start from assumptions, like a chess player—the consequences are inevitable and precise. The edifice of political ideas thus built up is wonderfully logical; in other words its intellectual integrity is unquestionable. Now, the Cartesian only admits one virtue; pure intellectual perfection. To all else he is deaf and blind. For him what is anterior and what is present are equally undeserving of the least respect. On the contrary, from the rational point of view, they assume a positively criminal aspect. He urges, therefore, the extermination of the offending growth and the immediate installation of his definitive social order. The ideal of the future, constructed by pure intellect, must supplant both past and present. This is the temper which produces revolutions. Rationalism applied to politics results in revolutionary doctrine and, *vice versa*, an epoch is not revolutionary unless it is rationalist. It cannot be revolutionary except in the proportion in which it is incapable of sensitiveness to history, incapable of perceiving both in past and present that other kind of reason which is not "pure" but vital.

The Constituent Assembly makes "solemn declaration of the rights of Man and of the Citizen" in order "that, it being possible to compare the acts of the legislative and executive powers, at any given moment, with the final aim of 'every' political institution, they may be the more respected, so that the demands of citizens, being founded henceforth on simple and unquestionable principles," etc., etc. We might be reading a geometrical treatise. The men of 1790 were not content with legislating for themselves: they not only decreed the "nullity" of the past and of the present, but they even suppressed future history as well, by decreeing the manner in which "every" political institution was to be constituted. To-day this attitude appears to us over arrogant. More, it appears to us to be narrow and crude. The world has become more complex and vast in our eyes. We are beginning to suspect that history, human life, cannot and "ought" not to be ruled by principle, like mathematical textbooks.*

It is illogical to guillotine a prince and replace him by a principle. The latter, no less than the former, places life under an absolute autocracy. And this is, precisely, an impossibility. Neither rationalist absolutism, which keeps reason but annihilates life, nor relativism, which keeps life but dissolves reason, are possibilities.

The sensibility of the age that is now beginning is characterised by its rejection of this dilemma. We cannot satisfactorily adjust ourselves to either of its terms.

* See Appendix I (*The Sunset of Revolution*).

CHAPTER IV

CULTURE AND LIFE

WE have seen that the problem of truth divided the men of the generations anterior to our own into two hostile schools of thought: relativism and rationalism. Each of them renounces what the other retains. Rationalism clings to truth and abandons life: relativism prefers the mobility of existence to the calm and immutability of truth. Our own spirit is alien from both of these positions: when we attempt to assume either of them we feel we are being mutilated. We see with perfect clarity the plausibilities of both, and we perceive, equally well, their complementary inadequacies. The fact that at other times the best minds had no difficulty in accommodating their inclinations to one or the other, according to their temperaments, indicates that they possessed a sensibility distinct from our own. We belong to an age in the proportion in which we feel capable of accepting its dilemma and arraying ourselves for battle on one side or the other of the trench it has dug. For the process of living is, in a very real sense, which will shortly come to light, an enlistment under certain standards and a preparation for battle. *Vivere militare est*, Seneca used to say, with the proud gesture of a legionary. What no one has any right to ask of us is, that we should take part in a struggle which we find has already been decided in our own hearts. For every generation must be what the Hebrews called *Neftali*, which means, "I have already fought my battles."

So far as we are concerned the old dispute has long ago been decided. We do not now understand how it is possible to speak of a human life in which the organ of truth has been amputated, or of a truth which requires the withdrawal of the vital stream before it can exist.

The problem of truth to which I have briefly alluded is only a single example. The same situation arises in the case of the moral and juristic standard which is supposed to regulate our wills, as truth regulates our thought. Goodness and justice, if they are what they claim to be, must necessarily be unique. Justice which is only just for a certain time, or for a certain race, cancels its own meaning. In ethics and law, then, too, the principles of relativism and rationalism arise, as they do also in art and religion. This is as much as to say that the problem of truth is dispersed throughout all the spiritual orders which we imply when we use the word, "culture."

Under this new name the question loses something of its technical aspect and approximates more to actual human energies. Let us therefore take it up at this point and try to examine it with all the severity we can, noting all its acutely dramatic character. Thought is a vital function, like digestion, or the circulation of the blood. The fact that the latter are processes active in space and among bodily tissues, while the former is not, makes no real difference so far as our particular theme is concerned. When the nineteenth century biologist refuses to consider as vital phenomena those which do not possess corporeal character he admits, at the start, a prejudice which is incompatible with any strict positivism. A doctor treating a patient has before him with equal immediacy the phenomenon of thought and that of respiration. An act of judgment is a tiny element of life: so too is an act of will. They are emanations or momenta of a self-centralised microcosm, the organic individual. I think my thoughts as I rearrange the chemical constituents of my food or pump blood from my heart. In all three cases we are concerned with vital necessities. To understand a biological phenomenon is to demonstrate its necessity for the preservation of the individual, or, what is the same thing, to discover its vital utility. My thought, therefore, finds its cause and justification in myself as an organic individual: it is an instrument for the benefit of my life, an organ of it, regulated and governed by it.

But, from another point of view, to think is to set before our individuality the nature of phenomena. The fact that we sometimes fall into error merely confirms the generally truthful character of thought. We give the name of error to a flawed thought, a thought that is not properly a thought at all. The business of thought is to reflect the world of phenomena, to adjust itself to them in one way or another: in short, to think is to think truth, just as to digest is to assimilate victuals. Error does not destroy the general truth of thought any more than indigestion annuls the fact of normal assimilative process.

Thought, then, has two distinct facets: on the one hand it comes into being as a vital need of the individual and is governed by the law of subjective utility: on the other hand it consists, actually, in an adaptation to phenomena and is directed by the objective law of truth.

It is the same with our volitions. An act of will is strictly centrifugal in character. It is an emanation of energy, an impulse which rises from the depths of the organism. The wish in its narrowest sense is always the wish to do something. The love of something and the simple desire that something may exist both intervene, no doubt, in the preparation of the act of will, but they are not the act itself. We wish, in the proper sense of the word, when, besides desiring that things may exist in a certain way, we determine to realise our desire and execute efficient acts which are to modify reality. In volitions there is a very clear manifestation of the vital urge of the individual. By their means he satisfies, revises and amplifies his organic needs.

Let us analyse an act of will at the point of the clearest emergence of its character. For example, take a case in which, after some vacillation and hesitation, we finally decide, at the end of a dramatic self-communion, to do something, and other possible resolves are repressed. We then observe that our decision has resulted from the fact that one of the competing proposals has appeared preferable in our eyes. Every wish, therefore, is constitutionally a wish to do the best that can be done in any situation, and this is an acceptance of an objective standard of what is good. Some will think that this objective volitional standard, this *summum bonum*, is the service of God; others will suppose that the best attitude consists in a circumspect egoism, or on the contrary, in doing the greatest good to the greatest number of their fellow-creatures. But, whatever its content, when a wish for something exists, the thing is wished for because it is believed to be the best, and we are only satisfied with ourselves, we have only wished fully and unreservedly, when we think we have adapted ourselves to a standard of the will which exists independently of ourselves, beyond our own individuality.

This double character, which we find in intellectual and volitional phenomena, is equally well authenticated when we come to æsthetic feeling or religious emotion. In other words a whole series of vital phenomena exists endowed with a doubly dynamic character, a strange duality. On the one hand they are the spontaneous product of the living person and have their cause and government centred within the organic individual: on the other hand they are bound by an inner necessity to submit to a government or law which is objective. Both aspects, it should be carefully observed, are mutually interdependent. I cannot think usefully, for my biological purposes, if I do not think what is true. Thought which would normally present us with a world divergent from the true world would lead us to commit incessant practical errors and consequently human life would by now have disappeared. In functioning intellectually, then, I do not succeed in self-adjustment, in being useful to myself, if I do not adjust myself to what is not myself, to my environment, to the transorganic world, to what transcends my own being. But also *vice versa:* truth does not exist if the person concerned does not think it, if there is not born in his organic being the mental act, with its inevitable aspect of intimate conviction. For thought to be true it must coincide with phenomena, with what transcends my own being; but at the same time, for this thought to exist, I must think it, I must believe in its truth, give it an intimate place in my life, make it immanent in the tiny biological globe which is myself.

Simmel, who has treated this question more acutely than anyone else, insists quite rightly on this peculiar aspect of the phenomenon of human life. The life of man, or the collection of phenomena integrated by the organic individual, possesses a transcendent dimension inasmuch as it abandons, so to speak, its own privacy and introduces itself into something alien, beyond its own limits. This dimension comprises thought, will, æsthetic feeling and religious emotion. I do not mean that in analysing, for example, the phenomenon of intelligence we accept the existence of the truth that it claims to contain. Although, as philosophers, we might not consider such existence substantiated, the phenomenon of thought does include, whether we like it or not, that claim; in fact, it consists of nothing else. So that when the relativist refuses to admit that a living being can think what is true he is convinced, in his capacity of living being, that there is truth in his very refusal.

Apart, then, from all theory, descending to simple facts and confining ourselves to the strictest positivism—which those who call themselves positivists never practise—we find that human life appears to be the phenomenon of the transcendence of the organism by certain activities which are immanent in it. Life, said Simmel, consists, precisely, in being something more than life; what is immanent in it is a transcendence beyond vital limits.

We can now give the word, culture, its exact significance. There are vital functions which obey objective laws, though they are, inasmuch as they are vital, subjective facts, within the organism; they exist, too, on condition of complying with the dictates of a *régime* independent of life itself. These are culture. The term should not, therefore, be allowed to retain any vagueness of content. Culture consists of certain biological activities, neither more nor less biological than digestion or locomotion. There was much talk during the nineteenth century, particularly in Germany, of culture as "spiritual life." The reflections we are now making permit us, fortunately, to give a precise sense to that expression, to the magic formula pronounced with gestures of such ecstatic rapture by the modern pretenders to sanctity. "Spiritual life" is nothing more nor less than the store of vital functions whose products or results possess a durability and importance independent of life. For instance, among the various ways in which we can behave to our neighbours our perception selects one in which it finds the special quality called "justice." The capacity for perceiving, for thinking justice, and preferring the just to the unjust is primarily a faculty with which the organism is endowed in order that it may promote its own peculiar and private convenience. If the sense of justice had been pernicious, or even superfluous, to the living being, it would have meant so heavy a biological burden that the human race would have succumbed. Justice, then, comes into existence in the character of a mere vital and subjective convenience; organically, juristic sensibility possesses neither more nor less value than the pancreatic secretion. However, justice, once it has been segregated by sentiment, acquires an independent value. It is an element in the actual concept of "the just" and includes the irresistible demand for its own existence. The requirements of what is just have to be fulfilled, though they may be inconvenient for life. Justice, truth, moral rectitude and beauty are things that are of value in themselves and not only in the proportion in which they are useful to life. Consequently, the vital functions in which these things are produced possess a value in themselves, apart from that of their biological utility. On the other hand the pancreas has no more importance that what is due to its organic utility. The secretion of such a substance is a function that ends with life itself. The value that justice and truth have in themselves, that plenary sufficiency in them that makes us prefer them to the very life that produces them, is the quality we denominate spirituality. In modern ideology "spirit" does not mean anything like "soul." The spiritual is not an incorporeal substance, not a reality. It is simply a quality that some things possess and others do not. That quality consists in the possession of a special significance and value.

The Greeks would have called modern spirituality "nous" but not "psyche" or soul. Accordingly, the perception of justice, the knowledge or thought of truth, artistic creation and enjoyment possess a significance of their own, a value in themselves, even if we ignore their utility to the living being which exercises such functions. They are spiritual life or culture. On the other hand secretions, locomotion and digestion are infra-spiritual life, purely biological life, with no significance or value outside the organism. For our clearer mutual understanding let us call such vital phenomena as do not transcend the biological by the name, spontaneous life.*

I do not think that the most fastidious devotee of culture and "spirituality" will find any of the special privileges attaching to these terms omitted from the foregoing definition. My only innovation has been to emphasise a certain aspect of them which the "culturalist" hypocritically attempts to obliterate and then ignores as though he had forgotten it. In fact, when we hear talk of "culture" and "spiritual life" there appears to be no question of anything but a separate life distinct from and out of all touch with the unfortunate and much despised "spontaneous" life. Anyone would suppose that thought, religious ecstasy and moral heroism can exist in the absence of the humble pancreatic secretion, the circulation of the blood and the nervous system. The culturist boards the adjective "spiritual" and cuts the cables that hold it to the substantive "life," *sensu stricto*, forgetting that the adjective is no more than a specification of the substantive and that without the latter the former could not exist. This is the fundamental error of rationalism in all its forms. The *"raison"* of the

* On identical grounds—and this is a most important point to note—spiritual activities are also primarily spontaneous life. The pure idea of science comes into being in the character of a spontaneous emanation of the person concerned, as a tear does.

rationalists, which pretends that it is not simply one vital function among many, and that it does not submit to the same organic regulation as the rest, has no existence: it is a foolish and purely fictitious abstraction.

There is no culture without life, there is no spirituality without vitality in the most literal sense that the word can bear. The spiritual is not less or more life than the non-spiritual.

CHAPTER V

THE DOUBLE IMPERATIVE

THE fact is that the phenomenon of human life has two faces, the biological and the spiritual, and is for that reason subject to two distinct forces which act on it in the manner of two poles attracting it in opposite directions. Thus, intellectual activity gravitates on the one side towards the centre of biological necessity and on the other is exposed to the intimations or rather positive orders of the extra-vital principle of logical law. Similarly, æsthetic feeling is in one direction subjective enjoyment, in the other beauty. The beauty of a painting does not consist in the fact, which is of no importance so far as the painting is concerned, that it causes us pleasure, but on the contrary we begin to think it a beautiful painting when we become conscious of the gently persistent demand it is making on us to feel pleasure.

The essential note in the new sensibility is actually the determination never in any way to forget that spiritual or cultural functions are equally and simultaneously biological functions. Further, that culture for that reason cannot be exclusively directed by its objective laws, or laws independent of life, but is at the same time subject to the laws of life. We are governed by two contrasted imperatives. Man as a living being must be good, orders the one, the cultural imperative: what is good must be human, must be lived and so compatible with and necessary to life, says the other imperative, the vital one. Giving a more generic expression to both, we shall reach the conception of the double mandate, life must be cultured, but culture is bound to be vital.

We are dealing, then, with two kinds of pressure, which mutually regulate and modify one another. Any fault in equilibrium in favour of one or the other involves, irremediably, a degeneration. Uncultured life is barbarism, devitalised culture is byzantinism.

There is a schematic and formalist way of thinking which proceeds without the co-operation of life and without direct intuition. It is a kind of cultural utopianism. We fall into it whenever we admit, without antecedent revision, certain principles of an intellectual, moral, political, æsthetic or religious nature, and, making the immediate assumption that they are valid, insist on the acceptance of their consequences. The present age is dangerously addicted to this unhealthy practice. The generations which invented positivism and rationalism made a most extensive synthesis of the questions raised by these systems of thought, the matter being of vital importance to them, and extracted their principles of culture from the strenuously elaborated scheme they propounded. In the same way, liberal and democratic ideas arose from practical contact with fundamental problems of society. But to-day hardly anyone proceeds on these lines. The characteristic fauna of modernity are the "naturalist" who swears by positivism without ever having taken the trouble to make a fresh synthesis of the theme formulated by that system, and the democrat who has never questioned the truth of democratic dogma. Hence results the farcical contradiction involved in the statement that actual European culture, while it pretends to be the only rational system, the only one, that is to say, founded upon practical reason, is really no longer lived and experienced in virtue of its rationality, but is embraced on mystical grounds. Pio Baroja's character, who believes in democracy as men believe in the Virgin of the Pillar,* is, like his predecessor the chemist Homais, a typical representative of modern times. The apparent predominance that has been acquired in this continent by retrograde forces does not arise from their being the bearers of principles superior to those of the opposition, but from the fact that they are at any rate free from such an essential contradiction and utter hypocrisy as that displayed by "rationalism." The traditionalist is in accord with his own inner being. His belief in mysticism is dictated by mystical motives. At any moment he can accept the ordeal of battle without being conscious of vacillation or reservation in his own mind. On the other hand anyone believing in rationalism as men believe in the Virgin of the Pillar can be said to have virtually ceased, in his heart of hearts, to believe in rationalism at all. Through mental inertia, through habit, through superstition, in a word, through traditionalism, he goes on adhering to the old rationalist theses which are now beyond the operation of creative reason and have become petrified, ritualised and byzantified. The rationalists of to-day perceive, in a more or less confused manner, that they are no longer in the right. Nor is this so much because they cannot now maintain a firm front against their adversaries, as because their home front has been demolished. The doctrines of liberty and democracy which they defend have ceased to satisfy the requirements of their champions; they do not fit with the necessary exactitude into their sensibility. This internal dualism robs the rationalists of the elasticity needed for controversy and they now enter the fray so torn

* An image of the Virgin profoundly reverenced in Spain particularly in Saragossa. (Translator's note.)

by domestic dissension that they are already half routed.

The extreme anomaly of such a situation obviously and necessarily requires the completion of the objective imperative by the subjective. It is not enough, for example, that a scientific or political concept should appear true for geometrical reasons, if we are to support it. It must also engender in us an absolute faith, free from all reservation. When this does not supervene it is our duty to retire from the concept and modify it as much as may be necessary, so that it may be made to agree strictly with our demands as organisms. A moral system which is geometrically perfect but leaves us cold and is no spur to action is subjectively immoral. The ethical ideal cannot content itself with being the most correct of ideals: it must also succeed in arousing our emotions. In the same way it is fatal for us to get into the habit of acknowledging as examples of consummate beauty works of art—for instance, the classics—which may be objectively of the greatest value but which do not induce enjoyment in us.

Consequently our activities require the regulation of a double series of imperatives, which might be summarised as follows:

ACTIVITIES		*IMPERATIVES*
Thought	*Cultural* {	*Vital* { Sincerity
Will	Truth / Goodness / Beauty	Emotional Drive
Sentiment		Enjoyment

The epoch which men have unfortunately decided to call "modern," viz., that beginning with the Renaissance and continuing up to our own day has been progressively dominated, more and more exclusively, by an altogether one-sided "culturalist" tendency. This one-sidedness involves a consequence of the gravest import. If we become so absorbed in adapting our convictions to what reason affirms to be truth we run the risk of believing that we believe, of feigning conviction because we are pleased to desire it. This prevents the realisation of culture in us and leaves it lying like a superficial pretence upon the substance of our effective life. In various degrees, but with a morbidly exasperated development during the last century, this has been the characteristic phenomenon of modern European history. People believed that they believed in culture: but strictly speaking they were dealing with a gigantic collective fiction which the individual did not apprehend because it had been forgery from the very foundations of his consciousness. On the one side marched principles, phrases and gestures, which were sometimes heroic: on the other the reality of existence, the life of every day and every hour. The English word "cant," signifying a scandalous duality between what people believe they are doing and what they are actually doing, does not really, as has been maintained, apply specifically to England only, but is common to all Europe. The Oriental, unaccustomed to the separation of culture and life, since he has always required vitality in the former, sees in Occidental behaviour a radical and all-embracing hypocrisy and cannot repress, when in contact with Europeans, a sentiment of disdain.

Such a dissociation between standards and their permanent translation into action would never have come about if we had been taught, together with the imperative of objectivity, that of self-consistency, which comprises the whole series of vital imperatives. It is necessary that at all times we should be sure that we do in fact believe what we presume we believe; that the ethical ideal we accept "officially" does in fact interest and stimulate the deeper energies of our personality. If we had been in the habit of so clarifying our inward situation from time to time, we should have automatically exercised due selection in culture and eliminated all such forms of it as are incompatible with life, utopian, and conducive to hypocrisy. On the other hand culture would not have been continually relegated to increasingly remote distances from the vitality which creates it, nor condemned at last, in a ghostly isolation, to petrifaction. So, in one of those phases of the drama of history, in which man needs all his vital resources to preserve himself from catastrophic circumstances and needs most of all those which are nourished and stimulated by faith in transcendental values, that is, in culture, it happens that, in such an hour as that which is now passing over Europe, everything fails him. And yet junctures like the present are the experimental test of cultures. Facts have brutally imposed on Europeans, through their own indiscretion, the immediate obligation to be self-consistent, to decide whether they authentically believe in what they believe, and they have discovered that they do not. They have called this discovery the "breakdown of culture." It is obvious that there is nothing of the sort: something had broken down long before, and that was the self-consistency of Europeans: the breakdown is that of their own vitality.

Culture arises from the basic life of the person concerned, and is, as I have pointed out with deliberate reiteration, life *sensu stricto*, that is, spontaneity, subjectivity. Little by little science, ethics, art, religious faith and juristic standards become separated from the person considering them and begin to acquire a consistency of their own, an independent value, prestige and authority. A time comes when life itself, the generator of all these conceptions, bows down before them, yields

to its own creation and enters its service. Culture has become objective and set itself up in opposition to the subjectivity which has engendered it. The words ob-ject, *ob-jectum*, *gegen-stand*, have the significance of that which is op-posed, that which establishes itself and sets itself up against the subject or person concerned as his law, precept and government. At this point culture comes to its fullest maturity. But certain limits have to be maintained to such an opposition to life, to such a separation between subject and object. Culture only survives while it continues to receive a constant flow of vitality from those who practise it. When this transfusion is interrupted and culture becomes more remote from life it soon dries up and becomes ritualised. Culture, then, has its hour of birth, which is its hour of lyric beauty, and its hour of petrifaction, which is its hour of ritualisation. There is culture in the bud and culture in flower.*
In ages of reform like our own culture in flower is bound to be suspected and emergent culture tended, or, what comes to the same thing, cultural imperatives are arrested and vital imperatives come into the foreground. Culture has to face the opposition of self-consistency, spontaneity and vitality.

* It is interesting to be present, through the medium of history, at this process, and see how what is later to be a simple principle of equity begins by being a magic rite, a legendary incantation, the special inclination of a group or simply a material convenience. It is always the same, in science, in morals, or in art. There should be a genealogy of culture.

CHAPTER VI

THE TWO IRONIES, OR, SOCRATES AND DON JUAN

HUMAN life is never without its two dimensions, culture and spontaneity; but only in Europe have these dimensions reached a point of complete differentiation, disengaging themselves to the point of forming two antagonistic poles. Neither in India nor in China have either science or morals ever achieved the status of powers independent of spontaneous life, and the exercise, in that capacity, of sovereignty over the latter. The thought of the Oriental, however acute and profound, has never separated itself from the thinker to the point of attaining to that definitely objective existence which a physical law, for instance, enjoys in European consciousness. It can be held that Oriental life is nearer perfection than Occidental; but its culture is evidently less culture than ours, realises less radically the sense that we give to the term. The glory, and perhaps the tragedy, of Europe is built up, on the contrary, upon the fact of Europeans having developed the transcendent dimension of life to its ultimate consequences. Oriental wisdom and morality have never lost their traditionalist character. The Chinese is incapable of conceiving an idea of the world by merely taking as his starting point the rationality, the truth of the idea itself. If he is to lend it his adherence and be convinced by it he has to see it as authorised by an immemorial past; that is to say, he must discover its foundations in the mental habits which his organism has absorbed from the race.

What exists by tradition is not what exists by culture. Traditionalism is really only a form of spontaneity. The men of 1789 exploded all the past and took pure reason as the basis of their formidable work of destruction; on the other hand, before the last Chinese revolution could be accomplished it had to be preached, and proved to have been suggested by the most orthodox Confucian dogma.

All the splendour and all the misery of European history are due, perhaps, to the utter divorce and antithesis that now subsist between the two opposing terms. Culture, or reason, has been refined to the last degree, almost to the point of severance from spontaneous life, which, for its own part, has remained equally isolated, but in a crude and practically aboriginal state. This condition of extreme tension inaugurated the unique dynamic quality, the endless vicissitudes and the permanent restlessness of the history of this continent. When we turn to Asiatic history we invariably seem to be watching the vegetal growth of a plant, of an inert being, without sufficient resilience to defy destiny. Vigorous resilience of this kind breaks out continually in the course of Occidental evolution, being caused by disturbance in the level of the two poles of life. Consequently, nothing illuminates the historical process of Europe to better effect than determination of the various relative positions taken up by culture and spontaneity.

For we must not forget that culture, or reason, has not always existed on this earth. There was a moment, the chronology of which is perfectly well known, at which the objective pole of life, viz., reason, was discovered. It may be said that on that day Europe, as such, came into being. Till then, existence on this continent had been merged with that in Asia or Egypt.

But one day, in the market-place at Athens, Socrates discovered reason.

I do not think that anyone can speak significantly of the duty of present-day man without having made himself thoroughly well acquainted with the meaning of the discovery of Socrates. It contains the key to European history and without it both our past and our present form an unintelligible hieroglyph.

Men had reasoned before Socrates; strictly speaking, two centuries of reasoning had already elapsed in the Hellenic world. But in order that something may be discovered it is obviously necessary that it should be already in existence. Parmenides and Heraclitus had reasoned, but they did not know it. Socrates was the first to realise that reason is a new universe, more perfect than and superior to that which we find, spontaneously, in our environment. Visible and tangible phenomena vary incessantly, appear and vanish, pass into one another: white blackens, water evaporates, man dies; what is greater in comparison with one thing turns out to be smaller in comparison with another. It is the same in the internal world of man: desires and projects change and contradict themselves; when pain lessens it becomes pleasure; when pleasure is repeated it grows wearisome or painful. Neither our environment nor our inner self affords a safe and solid refuge for the mind. On the other hand, pure ideas, or *logoi*, constitute a set of immutable beings, which are perfect and precise. The idea of whiteness contains nothing but "white"; movement never becomes static; "one" is always "one," just as two is always two. These ideas enter into mutual relation without ever discomposing one another or admitting vacillation: largeness is inexorably opposed to smallness; on the other hand, justice joins unity. Justice is, in fact, always one and the same.

It must have been with unparalleled emotion and enthusiasm that men saw, for the first time, the austere outlines of ideas, or "rationalities" rise before their minds. However mutually impenetrable two bodies may be, two ideas are much more so. Identity, for instance, resists, absolutely, any confusion with Difference. The virtuous man is always simultaneously more or less vicious; but Virtue is utterly distinct from Vice. Pure ideas are, then, clearer, more unmistakable, more impregnable, than the phenomena of our vital environment, and they behave in accordance with precise and invariable laws.

The feeling of excitement that the sudden revelation of this symbolical world provoked in the generations of the Socratic period comes home to us in the vibrant quality of the dialogues of Plato. There was no doubt about it: true reality had been discovered; and in contrast with it the other world, that presented to us by spontaneous life, underwent an automatic depreciation. This experience imposed on Socrates and his age a very definite attitude, indicating that the mission of mankind was to substitute the rational for the spontaneous. Accordingly, in the domain of intelligence, the individual felt constrained to suppress his spontaneous convictions, which were only "opinions" or *doxai*, and adopt, instead, the thoughts of pure reason, which were the only real type of knowledge or *episteme*. Similarly, in practical behaviour to deny and keep in abeyance all his native desires and propensities, so as to be able to afford a docile obedience to the mandates of rationality.

The theme of the Socratic period comprised, then, the design to get rid of spontaneous life and supplant it by pure reason. But this enterprise involves a duality in our existence, for spontaneity cannot be altogether annihilated; all that can be done is to arrest its progress as it grows, to restrain it and overlay it with the second type of life, that acting by reflex mechanism, in other words, rationality. In spite of Copernicus we continue to observe that the sun sets in the west; but this spontaneous evidence proffered us by our vision remains, so to speak, in abeyance, and has no consequences. We overspread it with the reflex conviction furnished to us by the pure reason of astronomy. Socratism, or rationalism, begets, on identical grounds, a double life in which our non-spontaneous character, or pure reason, is substituted for our true character, or spontaneity. It is in this sense that Socratic irony is used. For there is irony in every act by which we supplant a primary movement by a secondary, and instead of saying what we think, pretend to think what we say.

Rationalism is a gigantic attempt to destroy spontaneous life through irony, regarding it from the point of view of pure reason.

To what extremes can this process go? Can reason be self-supporting? Can it get rid of all the rest of life, the irrational, and go on living by itself? To this question no answer could be given at that time; the great attempt had first to be made. The shores of reason had been thoroughly explored, but its extent and content were as yet unknown. There were many centuries of fanatical rationalist investigation to come. Each new discovery of pure ideas increased faith in the unlimited possibilities of that gradually emerging world. The last centuries of Greece began the vast work. And scarcely had the Gothic invasion developed into a peaceful settlement in the west when the rationalist spark kindled by Socrates set fire to the new-born souls of France, Italy, England, Germany and Spain. Some centuries later, between the Renaissance and 1700, the great rationalist systems were constructed. With their aid pure reason invaded enormous dominions. Men were able, then, for a moment to embrace the illusion that the hope of Socrates was about to be realised, and that all life would finally accept the principles of pure intellect.

But in the very process of taking possession of the rational universe, and particularly on the morrow of the establishment of the triumphant systematisations of such men as Descartes, Spinoza and Leibniz, it was observed, with fresh astonishment, that the territory was limited. After 1700 strict rationalism begins to discover, not fresh types of reason, but the limits of reason, its boundaries marching with the infinite environment of the irrational. The century of critical philosophy supervenes, whose waves were to break so magnificently over the last hundred years and end by arriving, in our own day, at a definite demarcation of frontiers.

We now see clearly that Socrates and the centuries that succeeded him were in error, though their error has proved a fruitful one. Pure reason cannot supplant life: the culture of abstract intelligence is not, when compared with spontaneity, a further type of life which is self-supporting and can dispense with the first. It is only a tiny island afloat on the sea of primeval vitality. Far from being able to take the place of the latter, it must depend upon and be maintained by it, just as each one of the members of an organism derives its life from the entire structure.

This is the stage of European evolution which coincides with our own generation. The terms of the problem, after passing through a long cycle of changing positions, now stand at a point exactly opposite to that which they occupied in the mind of Socrates. Our age has made a discovery which is the inverse of his: he hit upon the direction which the power of reason takes; it has been left to us to discern its terminus. Our mission, therefore, is the contrary of his. Through rationalism we have once more discovered spontaneity.

This discovery does not mean a return to an ingenuous primevalism similar to that professed by Rousseau. Reason, culture *more geometrico*, is an acquisition we can never forgo. But it is necessary to correct Socratic or rationalist or culturalist mysticism, which is ignorant of the limits of reason, or fails to deduce honestly the consequences of such limitation. Reason is merely a form and function of life. Culture is a biological instrument and nothing more. When it is set up in opposition to life it represents a rebellion of the part against the whole. It must be reduced to its proper rank and duty.

The modern theme comprises the subjection of reason to vitality, its localisation within the biological scheme, and its surrender to spontaneity. In a few years it will seem ridiculous to have exacted from life an acquiescence in the service of culture. The mission of the new age is, precisely, the conversion of that relation and the demonstration that it is culture, reason, art and ethics that must enter the service of life.

Our attitude implies, then, a new irony, of a type inverse to that of Socrates. While he mistrusted spontaneity and regarded it through the spectacles of rational standards, the man of the present day mistrusts reason and criticises it through the spectacles of spontaneity. He does not deny reason, but rejects and ridicules its pretensions to absolute sovereignty. Old-fashioned people may perhaps consider this disrespectful. In any case it is inevitable. The hour at which life will present its demands to culture can no longer be postponed. "Everything which, at the moment, we call culture, education and civilisation, will have to appear, one day, before the infallible judgment seat of Dionysus," declared Nietzsche, prophetically, in one of his early works.

Such is the irreverent irony of Don Juan, the enigmatic figure which our age has continued to prune and polish to the point of finally bestowing a precise significance upon it. Don Juan revolts against morality because morality had previously risen in rebellion against life. Only when a system of ethics is current which affirms plenary vitality as its first rule, will Don Juan agree to submit. But this will mean the succession of a new type of culture, the biological. Pure reason has, then, to surrender its authority to vital reason.

CHAPTER VII

VALUATIONS OF LIFE

WHEN we say that the essential theme of modern times, and the mission of the present generations, is constituted by an energetic attempt to regulate the world from the point of view of life, we run a serious risk of this allegation not being properly understood. For it may be considered that the attempt has often been made already; and further, that the vital point of view is that which is innate and primary in mankind. Surely this is the procedure of the savage, the pre-cultural man?

Not at all. The savage does not regulate the universe, exterior and interior, from the point of view of life. To take up a certain point of view implies the adoption of a contemplative, theoretical or rational attitude. We might just as well substitute the word, principle, for the expression, point of view. Now, there is nothing more hostile to biological spontaneity, to pure living, than the search for a principle from which we may derive our thoughts and actions. The selection of a point of view is the initial action of culture. Consequently the vital imperative which dominates the destiny of the coming race bears no relation to the return to a primitive manner of existence.

What we are dealing with is a fresh manifestation of culture; a consecration of life, which has hitherto been a bare fact, and, so to speak, a cosmic accident; this consecration is converting it into a principle and a right.

It may seem a matter for surprise when we come to think of it, but it is a fact, that while life has promoted the most various entities to the rank of principles it has never tried to make itself a principle. Life has proceeded under the guidance of religion, science, morality and economics; it has even proceeded under the capricious direction of art or pleasure; the one expedient that has never been essayed is that of living intentionally under the guidance of life. Fortunately, mankind has always more or less lived in this way, but such living has been unintentional; as soon as men saw what they were doing they repented, and experienced a mysterious remorse.

This phenomenon in human history is too remarkable not to receive some measure of attention.

The reason for our promotion of any entity to the dignity of a principle is to be found in our discovery of some superior merit in it. Because we consider that it is worth more than other things we have a preference for it and contrive to give it precedence over them. In addition to the real elements which compose the nature of an object the latter possesses a series of unreal elements which constitute its value. Canvas, lines, colours and forms are the real ingredients of a painting; beauty, harmony, grace and simplicity are its values. A thing, therefore, is not a value in itself, but it has values, is valuable, in fact. And these values, which reside in things, are qualities of unreal type. The lines of the painting are seen, but not its beauty; its beauty is "felt" or assessed. Assessment is to values what sight is to colours and hearing to sounds.

Every object enjoys, on identical grounds, a kind of dual existence. On the one hand it is a structure of real qualities, which we can perceive; on the other it is a structure of values, which are only apparent to our assessive capacity. And in the same way as there is a progressive experience of the properties of things—we discover, to-day, aspects and details which we did not see yesterday—there is also an experience of their values, a succession of discoveries of them, a greater subtlety in their assessment. These two experiences, the sensuous and the assessive, proceed independently of each other. Sometimes a thing is perfectly well known to us as regards its real elements, yet we are blind to its values. The paintings of El Greco hung for more than two centuries on the walls of the courts of justice, churches and galleries. Yet up to the second half of the last century their specific values had not been discovered. What had formerly seemed faulty in them was suddenly revealed as the repository of the highest æsthetic qualities. The assessive faculty, which makes us "see" values, is therefore completely distinct from sensuous or intellectual perspicacity. And there are men of genius in the domain of assessment as there are in that of thought. When Jesus, by enduring a blow without resentment, discovered humility, he enriched our assessive experience with a new value. In the same way, before Manet, no one had perceived the charm that lies in the trifling circumstance that the life of phenomena proceeds in the envelope of vague luminosity provided by the air. The beauty of "*plein air*" painting has made a definite contribution to the store of æsthetic values.

If the nature of values is analysed a little further, we shall find that they possess certain characters, alien from real qualities. It is essential, for instance, for every value to be positive or negative: there is no middle term. Justice is a positive value: the acts of perceiving it and esteeming it are identical. Injustice, on the other hand, is also a value, but a negative one: our perception of it is actually a condemnation of it. Moreover, every positive value is always superior, equivalent or inferior to other values. When we have a clear intuition of any two, we observe that one exceeds the other and is set above it, each remaining in a different rank. The elegance of a dress compels our esteem because it is a positive value; but if we compare it with the honesty of a personal character, we see that the former, without losing its estimable character, is nevertheless subordinate to the latter. Honesty is worth more than elegance, it is a superior value. For this reason we esteem both, but prefer honesty. This mysterious mental activity which we call "preference" proves that values constitute a strict hierarchy of fixed and immutable ranks. We may be mistaken in our preference in any given case, and place the inferior above the superior, just as we may make a mistake in calculations without that circumstance destroying the strict validity of arithmetical computation. When any error in preferential judgment becomes constitutional in a person, in an age or in a nation, and the inferior comes to be placed habitually above the superior, thus disturbing the objective hierarchy of values, it is a perversion, an assessive malady, with which we have to deal.

The foregoing brief remarks on the world of values were inevitable if we were to make the fact intelligible that up to the present time life has not been consecrated as a principle capable of regulating, in its turn, the rest of the phenomena of the universe. It now becomes possible to divine whence an explanation of this extraordinary circumstance may be obtained. Can it be that the specific vital values have not yet been discovered? And is there not some reason for the delay in such discovery?

It is extremely instructive to cast a glance, though a very summary one, at the different valuations that have been made of life. It will be sufficient for our most pressing needs to turn our attention to one or two of the outstanding peaks in the process of history.

Asiatic life culminates in Buddhism: this is the classical type, the ripe fruit, of the Oriental tree. In Buddhism the Asiatic soul expresses its radical tendencies with the clarity, simplicity and fullness characteristic of all classicism. And what is life according to Buddha?

The acute perception of Gautama hit upon the essence of the vital process and defined it as a thirst-*tanha*. Life is thirst, ardour, solicitude, desire. It is not attainment, for that which is attained is automatically converted into a starting point for some new desire. Existence regarded in this way, as an overwhelming and insatiable thirst, seems purely evil, and has no more than an absolutely negative value. The only reasonable attitude to it is to reject it. If Buddha had not believed in the traditional doctrine of reincarnation, his only dogma would have been that of suicide. But death does not annul life: the individual migrates in person through successive existences, a prisoner of the eternally and senselessly revolving wheel driven by cosmic thirst. How is one to escape life, how frustrate the endless chain of rebirth? This is all that ought to occupy our attention, all that can have value in life: flight, the evasion of existence, annihilation. The *summum bonum*, the supreme value opposed by the East to the *summum malum* of life, is, precisely, not-life, the pure not-being of the individual.

It should be noted that Asiatic sensibility is, at bottom, of a type inverse to that of Europe. While the latter conceives happiness as fully developed life, as the life of life to its completest extent, the most vital solicitude of the Indian is to cease to live, to efface himself from life, to sink into an "infinite inane," to cease to be conscious of himself. The initiate says: "Just as the enormous seas of the world have but one savour, the savour of salt, so the whole of the Law has but one savour, the savour of Salvation." This salvation is simply extinction, *nirvana*, *parinirvana*. Buddhism furnishes a technique for the acquisition of such a state and he who practises its precepts is enabled to give life a sense that it does not naturally bear: he converts it into an instrument for the annulment of its own being.* The life of the Buddhist is a "path," a road to the annihilation of life. Gautama was the "Master of the Path," the guide upon the highways to *Nihil*.†

While the Buddhist starts with an analysis of life which results in a negative valuation of it, and then discovers his *summum bonum* in annihilation, the Christian does not assume, to begin with, an assessive attitude to earthly existence at all. I mean that Christianity does not start with meditations upon life itself, but that it commences at once with the revelation of a supreme reality, the divine essence, which is the meeting point of all

---

* The stages in such a life of annihilation indicate also the various grades of sanctity. The ancient canon is divided into four principal ranks:

(*a*) The *Srotanpana*, literally, "He who has reached the river," that is to say, he who has set his foot on the path of the Law and thus begun his task of self-salvation.

(*b*) The *Sakrdagamin*, "He who returns but once again": this grade is occupied by one who has succeeded in annulling his desires and passions, but still retains a final remnant of them and is therefore obliged to be reborn once more into this world.

(*c*) The *Anagamin*, "He who returns no more," i.e., is not reborn on earth, but does exist again, for one further period, in the world of the Gods.

(*d*) The *Arhat*, the highest grade, only attainable by the monk, and in which the extinction of *nirvana* is fully achieved. See Pischel, *Leben und Lehre des Buddha*, pp. 87-88 (1921).

† A more detailed exposition of the Buddhist Law would have to draw attention to the fact that *nirvana* does not consist, from the Oriental point of view, in an absolute Nothing. It is really the annulment of personal existence, and therefore, for a European, equivalent to complete non-existence. But it is characteristic of Asiatic thought that the Oriental has a positive idea of existence that transcends the personal.

types of perfection. The infinity of this *summum bonum* reduces all possible others to negligible quantities. Accordingly, "this life" is of no value, good or bad. The Christian is not, like Buddha, a pessimist, but neither is he, strictly speaking, an optimist as regards earthly things. The one value recognisable by man is the possession of God, the beatitude which is only attainable beyond this life in a future existence, the "other life" or the "blessed life."

The valuation of earthly existence begins, for the Christian, when such existence is brought into relation with beatitude. That which is in itself indifferent and devoid of all intrinsic or immanent value of its own can then be converted into a great good or a great evil. If we assess life on the basis of what it actually is, if we affirm it for its own sake, we desert God, who is the one true value. In that case life is an incalculable evil, absolute sin. For the essence of all sin consists, for the Christian, in the application of a worldly standard of judgment to our behaviour. Now, desire and pleasure imply a tacit and profound acquiescence in life. Pleasure, as Nietzsche said, "longs for eternity, longs for deep, deep eternity," its aspiration is to perpetuate the moment of delight and it cries "*da capo*" to the reality which charms it. Accordingly Christianity makes the desire of pleasure, *cupiditas*, its capital sin.* If, on the contrary, we deny life all intrinsic value and maintain that it only acquires justification, sense and dignity when it is given an intermediate status and made a time of testing and practical training for the attainment of the "other life," then we invest it with a highly estimable character.

The value of existence lies, then, for the Christian, in something outside its own limits. Not in its own nature, but

---

*Habes apostolum dicentem radicem omnium malorum ease cupiditas. (*Saint Augustine.*)

beyond its horizon; not in its own immanent qualities, but only in the transcendent and ultra-vital value that belongs to beatitude can life achieve any considerable dignity.

Temporal phenomena are mean and shallow streams of misfortune that acquire nobility only when they widen into the sea of eternity. This life is only good in so far as it is a medium for progress and adaptation to the other. Instead of living for its own sake man should transform it into a preparatory exercise and a continuous training for death, whose hour is the commencement for the only true life. Training is perhaps the contemporary word which best translates what Christianity calls asceticism.

In the arena of the Middle Ages was fought out, with gallantry on both sides, the battle between the vital enthusiasm of the Goth and the Christian disdain for life. Those feudal lords, in whose youthful organisms the primitive instincts ramped like wild beasts in their cages, gradually surrendered their indomitable zoological violence to the ascetic discipline of the new religion. They were used to feeding on bears' meat and the flesh of deer and wild boar. As a consequence of this diet they had to be bled every month. The process of hygienic bleeding, which prevented the occurrence of a physiological explosion in the patient, was called "minutio." Well, Christianity was the integrating "minutio" of the biological excess the Goth brought with him from his native forests.

Modern times represent a crusade against Christianity. Science and reason have gradually demolished that celestial future world which had been erected by Christianity at the frontier beyond the grave. By the middle of the eighteenth century the divine world to come had evaporated. This life was all that remained to man. It seems as though we have now come to a time in which vital values are at last about to be revealed. Yet the revelation is still to seek. The thought of the last centuries, though anti-Christian, is seen to have adopted an attitude in regard to life which has a strong resemblance to that of Christianity. What are the substantive values for the modern man? Science, art, morals, justice: what has been called culture. Are these not vital activities? Certainly they are, and in that sense we might suppose, for a moment, that modernity had succeeded in discovering the immanent values in life. But a little further analysis shows us that this interpretation is not an exact one.

Science is the faculty of the understanding that pursues truth through the medium of truth itself. It is not the biological function of the intellect which, like all other vital powers, is the servant of the whole organism of the living being and derives its regulation and modulation from that organism. Precisely in the same way the sentiment of justice, and the actions to which that sentiment gives rise, originate in the individual but do not refer back to him as their centre; their final relation is to the extra-vital value of justice itself. The formula, *pereat mundus fiat justitia*, expresses, with the fury of an extreme radicalism, disdain for life and the modern apotheosis of cultural standards. Culture, the supreme value worshipped by the two positivist centuries, is also an ultra vital entity, which occupies, in modern estimation, exactly the same position as beatitude formerly enjoyed. The European of yesterday and the day before yesterday has no conception, any more than has the Oriental, of a life of immanent values, such as may properly be called vital.

The "Good, the Beautiful, the True" only achieve estimable importance in the service of culture. The doctrine of culture is a kind of Christianity without God. The attributes of the latter sovereign reality— Goodness, Truth and Beauty—have been amputated or dismantled from the divine person, and once they were separated they became deified. Science, Law, Morality, Art, etc., are activities which were originally vital, magnificent and spirited emanations of life, which the culturalist only appreciates in so far as they have been antecedently disintegrated from the integral process of vitality which creates and sustains them. The life of culture is habitually called a life of the spirit. There is no great distinction between the latter and the "blessed life." Strictly speaking, the one cannot claim a larger share of immanence than the other in actual historical fact, which is always life. Upon investigation it is very soon apparent that culture is never a fact or an actuality. The movement in the direction of truth or the theoretical exercise of the intelligence is certainly a phenomenon whose existence can be verified in different forms to-day, just as it could be yesterday or at any other time, no less than the phenomena of respiration or digestion. But science, or the possession of truth, is, like the possession of God, an event that neither has happened nor can happen in "this life." Science is only an ideal. The science of to-day corrects that of yesterday, and that of to-morrow corrects that of to-day. Science is not a fact which is brought about in time: as Kant and his whole age thought, complete science or true justice are only produced in the infinite process of infinite history. Hence culturalism has always an extremely "progressive" character. The meaning and value of life, which is essentially present actuality, are for ever awaking to a more enlightened dawn, and so it goes on. Real existence remains perpetually on the subordinate level of a mere transition towards an utopian future. The doctrines of culture, progress, futurism and utopianism are a single unique ism. Under one denomination or the other we invariably find the attitude of mind in which life for its own sake is a matter of indifference, and only acquires value if it is considered as an instrument or as a basis for the use of a culture operating in the "Beyond."

To what point it is illusory to desire to isolate from life certain organic functions to which the mystic name of spiritual is given we have learned only too well during the recent evolution of Germany. Just as the Frenchman of the eighteenth century was "progressive," the German of the nineteenth has been culturalist. All the best German thought from Kant to 1900 can be subsumed under the rubric, Philosophy of Culture. We should scarcely be able to enter upon it before we perceived its resemblance, in form, to medieval theology. There has only been a substitution of certain new entities for the old: where the ancient Christian thinker said, God, the contemporary German says, Concept (Hegel), Supremacy of Practical Reason (Kant, Fichte), or Culture (Cohen, Windelband, Rickert). The illusory deification of certain vital energies at the cost of all the rest, the disintegration of what can only exist in composition, e.g., science and respiration, morals and sexuality, justice and a sound secretional system, bring in their train the great organic disasters, the gigantic catastrophes of thought. Life imposes on all its activities an imperative of integration, and whoever says "yes" to one must affirm all.

Is it not an alluring idea to reverse the present attitude completely and instead of looking outside life for its meaning to turn our attention to life itself? Is it not a theme worthy of a generation which stands at the most radical crisis of modern history, if an attempt be made to oppose the tradition and see what happens if instead of saying, "life for the sake of culture," we say, "culture for the sake of life"?

CHAPTER VIII

VITAL VALUES

WE have seen that whenever, in all previous cultures, an attempt was made to discover the value of life or, in the current phrase, its "meaning" or justification, application was made to conceptions that lie beyond its limits. The value of life always seemed to consist in something transcending it, for the achievement of which life was merely an avenue or an instrument. Of itself, in its immanent aspect, it appeared quite devoid of estimable qualities, when, indeed, it was not considered to be charged exclusively with negative values.

The reason for this persistent phenomenon is not in doubt. For does not the business of living consist, precisely, in giving one's attention to what is not life? To see is not to contemplate one's own ocular apparatus, but to unveil the world about us, to allow oneself to be overwhelmed by the impressive flood of cosmic form. Desire, the vital function which best symbolises the essence of all the rest, is a constant mobilisation of our being in directions that lead beyond it: it resembles a tireless archer, despatching us endlessly to the targets that excite our emotions. In the same way thought always thinks something that is not itself. Even in the case of reflection, when we do think of our own thoughts, the latter are bound to have an object which, again, is not thought.

It was an incalculably disastrous error to maintain

that life, when left to its own devices, tends to egoism, for in its root and essence it is indisputably altruistic.

Life is the cosmic realisation of altruism, and exists solely as a perpetual emigration of the vital Ego in the direction of the Not-self.

This transitive character of vitality has not been neglected by the philosophers who have investigated the value of life. Observing that people could not live without taking an interest in one thing or another, they concluded that it was really those things which were interesting, and not the fact itself of being interested. A similar equivocation would be committed in the supposition that what was valuable in the practice of climbing was the mountain peak and not the ascent. In meditating on life one has to evade it, to leave all its interior movements in suspense and ineffective, and contemplate its flow from without, just as the turbulent race of a torrent can be witnessed from the bank of a river. For this reason Fichte very properly declared that to philosophize is, in its true meaning, not to live, and that to live, in its true meaning, is not to philosophize. All men, and we ourselves, when we live our spontaneous life, toil in the service of science, art or justice. Within our vital mechanism these are the ideas that stimulate our activity, these are what has value "for" life. But when existence is regarded from a point outside itself we can see that these fine things are only pretexts invented by vitality for its own use, just as an archer seeks a target for his arrow. It is not, therefore, transcendent values which give a meaning to life but, on the contrary, the admirable generosity of spirit in the latter, which always requires something alien to itself to kindle its enthusiasm. I do not mean by this that all these great stimulating ideas have a merely fictitious value: my only object is to point out that there is no less value than theirs in the capacity, which constitutes the essence of life, to be stirred by what is estimable.

It is therefore necessary, in philosophy, to accustom oneself to keep one's attention fixed upon life itself, without permitting oneself to be carried away by its movement towards the ultra-vital. Life is like crystal, the transparent medium through which we can see other objects. If we permit ourselves to be deluded by the strong desire that any transparent thing implants in us, to pass heedlessly through it to something on the other side, we shall never see the crystal. In order to reach the point of perceiving it we have to disregard everything behind the glass and bring our eyes back to itself, to that ironical substance which seems to have a self-annihilating quality and to permit itself to be penetrated by what lies beyond it.

An effort similar to that of the above-mentioned ocular adjustment has to be made if we are to observe life instead of attaching ourselves to it and identifying ourselves with its impulses. We then discover the values which are peculiar to it.

The first of these is closely connected with life regarded as *genus*, whatever may be its direction and content. It is enough to compare the mode of existence in the mineral kingdom with that proper to all living organisms, the former being the simplest and most primitive, to obtain a clear intuition of this particular value of life. Whenever we perceive an indubitable difference of rank between two things, whenever, in concentrating our attention upon them, we become aware that there is a spontaneous subordination of one to the other, producing a hierarchy, we "see" their values. And, in point of fact, upon comparison of the most painful and sordid life with the most perfect of stones we instantly become aware of the superior dignity of the former. So evident is this superiority of the act of living to all that is not life that neither Buddhism nor Christianity has been able to deny it. I believe I have already indicated that the *nirvana* of the Indian is not, in any strict sense, the mere annihilation of life; in other words, it is not absolute death. There exist in the Asiatic conception of the world—perhaps it is the most characteristic feature of Eastern thought—two forms of existence and of life: the individual, in which the living being is conscious of himself as a part distinct from the whole, and the universal, in which he is everything, and therefore nothing in particular. *Nirvana*, reduced to its simplest terms, is the dissolution of the individual life in the great living sea of the universe; it therefore preserves the generic character of vitality which, in our Western view, the stone lacks. Similarly, what Christianity prefers to this life is not inanimate existence, but precisely that "other" life, which may be as much "other" as you please, but which coincides with "this life" in its principle, i.e., in being life. Bliss, in the theological sense, has distinct biological features, and on the day, not perhaps so far distant as the reader thinks, when a general science of biology is constructed, in which current biology will be only a section, the fauna and physiology of heaven will be defined and studied biologically, as comprising one of so many "possible" forms of life.

Life, then, does not require to possess any fixed content—of asceticism or culture—before it can have value and meaning. Life is valuable, no less than justice, beauty or beatitude, for its own sake. Goethe was perhaps the first man to have a clear notion of this idea when he said, surveying his entire existence: "The more I think of it the more evident it appears to me that life exists simply for the purpose of being lived." This self-sufficiency of vitality in the sphere of valuations frees it from the dependence to which it has erroneously been relegated, and from which arose the doctrine that the act of living was only estimable when it was employed in the service of something else.

The fact is that on the actual plane of life itself, when measurements are taken, as from a sea-level, from its hierarchical altitude, forms of the act of living, all more or less valuable, can be distinguished.

In this connection Nietzsche was the foremost of all seers. We owe to him the discovery of one of the most fruitful thoughts that have fallen into the lap of our age. I refer to his distinction between ascendent and descendent life, between life as a success and life as a failure.

There is no necessity to have recourse to extra-vital considerations, theological, cultural, etc. Life itself selects and constructs its hierarchy of values. Let us imagine that we have before us a collection of specimens of a single zoological subdivision of some kind; for instance, that of the horse. Even if we reject all utilitarian points of view, we can still range the specimens in a graduated series in which each animal represents an evolutionary stage in the realisation of equine possibilities. Seen from the one end the series presents life in its ascendent aspect, that is, life becoming, on each occasion, more and more life: seen from the other it enables us to note the progressive descent of vitality to the stage at which degeneration of type sets in. Moreover, between one extreme and the other we shall be able to determine accurately the point at which vital form begins to travel definitely towards perfection or decadence. From that point downwards we consider the specimens "poor": for in them the biological potency of the type becomes impoverished. On the other hand, from that point upwards there is a gradual evolution of the "pure-blooded" class, of the "noble" animal, in which the type achieves "nobility." There are here two values, the one positive, the other negative, both purely vital, viz., nobility and poverty. In both classes strictly zoological activities come into operation, viz., health, strength, speed, mettle and an organically well-proportioned form, or else the decline and disappearance of these attributes. This perspective of purely vital estimation of values does not, of course, exclude mankind. It is high time to make an end of the traditional hypocrisy which pretends it cannot see in certain human individuals, culturally of little or no interest, a splendour and grace of an animal type. I mean, of course, that grace of animal type which is peculiar to human beings, the grace of the *genus* "man" in its exclusively zoological aspect, but with all its specific potencies developed to which, strictly speaking, no culture can make any addition. (Culture is merely a special direction which we give to the cultivation of our animal potencies.) The most remarkable case of this kind is that of Napoleon, before whose dazzling perfection of vitality the saintly adepts of both schools, both the mystic and the democrat, are wont to veil their gaze.

It is extraordinary how difficult some people find it to accept the inevitable duplicity with which reality often comes before us. Their difficulty is due to their only wanting to retain one view of things and to their denial or deliberate concealment of the view which contradicts it. Ethically and legally Napoleon may have been a bandit—a proposition, by the way, not so easy to prove unless the demonstrator first takes care to enrol himself in some definite school of thought—but, in any case, and whether we like it or not, it is indisputable that in him the whole structure of man vibrated to its depths, for he was, as Nietzsche said, "The bow strung to the highest possible tension." It is not only the cultural and objective value of truth which is the measure of intelligence. When the latter is regarded as a pure vital attribute we call its peculiar virtue dexterity, bearing in mind that what makes speed really estimable in a horse is not actually the fact that we use that quality to arrive quickly at a preordained spot.

There can be no doubt that life in the antique world was less affected by trans-vital values, religious or cultural, than life as inaugurated by Christianity and its modern developments. A good Greek and a good Roman are nearer the zoological buff than a Christian or a "progressive" of our own day. And yet Saint Augustine, long under pagan influence and long accustomed to an "antique" view of the world, was unable to rid himself of a profound respect for the "animal" values of Greece and Rome. In the light of his new faith an existence without God must have seemed worthless and empty to him. Nevertheless, the clarity with which the vital grace of paganism manifested itself to his intuitive faculty was such that he was wont to express his respect for it in the equivocal phrase, *Virtutes ethnicorum splendida vitia*, "The virtues of the pagans are splendid vices." Vices? Well, then, they are negative values. Splendid? Well, then, they are positive ones. This contradictory valuation is the utmost that life has in the past been able to acquire. The impressive grace of life forces itself upon our sensibility; but at the same time our appreciation has a savour of delinquency about it. Why is it not a delinquency to say that the sun gives light, while on the other hand it is delinquent to think that life is splendid, that it makes its voyage laden to the brim with a wealth of values, just as the galleys of Ophir were rowed to their ports bearing cargoes of pearls? To overcome this inveterate hypocrisy in the face of life is perhaps the lofty mission assigned to modernity.

CHAPTER IX

SIGNS OF THE TIMES

THE discovery of immanent values in life by Goethe and Nietzsche was an intuition of genius which anticipated a future event of the most transcendent importance: the discovery of the same values by the sensibility possessed in common by a whole epoch. This duly foreseen epoch, prophesied by the seers of genius I have just mentioned, has now arrived: it is our own.

The trouble that certain people may take to ignore the serious crisis through which Western history is passing to-day will be in vain. The symptoms are only too evident, and the most obstinate renegade is secretly and continuously conscious of them in his own heart. Little by little ever larger areas of European society become the field of a strange phenomenon which might be called "vital disorientation."

We possess orientation when there does not exist in our minds the least doubt of the positions of north and south, the ultimate goals which serve the purpose of ideal indicative points for the guidance of our faculty of action and of our movements. Since life is so essentially action and movement, the system of goals towards which our acts are despatched and towards which our movements advance plays an integrating part in the living organism. The things to which we aspire, the things we believe in, the things we venerate and adore, have been created in the environment of our individuality by our actual organic potency and constitute a kind of biological drapery in which we are indissolubly swathed, body and soul. Our life proceeds as a function of our environment, which in its turn depends upon our sensibility. The world of the spider is not the same as that of the tiger or as that of mankind. The world of an Asiatic is not the same as that of a Greek of Socrates' time or as that of one of our own contemporaries.

This means that as the living being evolves his environment is proportionately modified and that there is, above all, a proportionate variation in the perspective that his enviroment offers. Let us imagine a moment of transition during which the great goals that yesterday furnished our landscape with so definite an architecture have been deprived of their lustre, of their attractive power and of their authority over us, while at the same time those that are destined to replace them have not yet acquired complete clarity of outline and competent vigour of growth. At such a season the landscape in the neighbourhood of the observer seems to break up, vacillate and quake in all directions; his steps, too, will be vacillating, for his cardinal points are oscillating and becoming obliterated and the very roads beneath his feet are melting away in serpentine undulations as though in flight before him.

Such is the situation with which European existence is confronted to-day. The system of values by which its activity was regulated thirty years ago has lost its convincing character, its attractive force and its imperative vigour. The man of the West is undergoing a process of radical disorientation because he no longer knows by what stars he is to guide his life.

Let us be accurate: thirty years ago the immense majority of European humanity were still living for the sake of culture. Science, art and justice were considered to be self-sufficient: a life that placed itself entirely at their disposition had a clear conscience. No one questioned the adequacy of those ultimate repositories of prestige. The individual, certainly, could ignore them and devote himself to other less stable interests; but in the very act of so doing he would recognise that he was yielding to the licentious freedom of a caprice, below the surface of which the cultural justification of existence continued unshaken. He was conscious that he might return at any moment to the canonical and securely established form of life. In the same way the sinner of the Christian era in Europe used to regard his own sinful life as afloat upon the ocean of the profound and living faith in the laws of God which filled the recesses of his soul.

During the period covering the end of the nineteenth century and the beginning of the twentieth the politician who invoked "social justice," "public liberty," and the "sovereignty of the people" at his meetings was sure of a sincere and effective response to these conceptions in the intimate sensibility of his audience. So was the hieratically solemn apologist of the human dignity of art. To-day this is not the case. Why not? Have we ceased to believe in these great things? Do we take no further interest in justice, science and art?

There can be no doubt about the answer to these questions. We do still believe, but in a different way, and as though we were posted at a different spatial interval. Perhaps the example that best exhibits the measure of the new sensibility is to be found in the art now being produced by the young. With surprising unanimity the most recent of the generations of all Western countries is creating an art, musical, pictorial and poetic, which is infuriating the men of the generations anterior to their own. Even people of middle age, who are more sympathetically inclined, cannot bring themselves to appreciate the new art, for the simple reason that they are unable to understand it. It is not that they think it better or worse than the art of the past; they do not consider it art at all, and consequently they quite sincerely believe that they are dealing with a gigantic fraud which has been allowed to spread its ramifications over the whole of Europe and America.

It is not difficult to account for the impassable gulf separating the opinions of old and young as regards the art of the present day. In previous stages of artistic evolution variations of style, which were sometimes profound—one remembers the changes brought about in romanticism through its conflict with neo-classicism—were always limited to alterations in the objects of aesthetic feeling and substitutions of one for another. The forms of beauty preferred at various times were different. But throughout these variations in the objects of aesthetic feeling there remained invariable the attitude of the aesthete and the spatial interval between himself and the object. In the case of the generation which is now reaching maturity the transformation is much more radical. The art of the young does not differ from traditional art so much in its objects as in its radical change of subjective attitude to art itself. The general symptom of the new style, evident in all its multiform manifestations, is to be found in the circumstance that art has been dislodged from its position in the "serious" zone of life, has, in fact, ceased to be a centre of vital gravitation. The semi-religious character, cultivating pathos of a sublime type, which aesthetic taste has been acquiring for two centuries, has now been completely extirpated. Art, in the consciousness of the new race, becomes philistinism or not-art as soon as it is taken seriously. The "serious" is the central zone through which the axis of our existence passes. But art is incapable of supporting the entire weight of life. When it tries to do so it breaks down and loses its essential grace. If, on the other hand, we displace our aesthetic attention and transfer it from the centre of life to the circumference, and if instead of taking art seriously we take it for what it is, an entertainment, a game, a diversion, then the work of art will once more assume the lyric charm with which it has been associated in the past.

The disagreement between old and young on aesthetic questions is therefore too radical to admit of the possibility of amelioration. So far as the old are concerned the lack of seriousness in the new art is a defect which is quite enough to render it negligible: while in the view of the young that very lack of seriousness counts as the supreme value of art, and accordingly they do their best, by determined and deliberate cultivation, to achieve it.

This revolutionary attitude to art reveals one of the most widespread features in the new reaction to existence: it is what I long ago called the sense of life as a sport and as a festivity. Cultural progressivity, which has been the religion of the last two centuries, could not assess the activities of mankind except with an eye to their results. The necessity and obligations of culture impose on humanity the execution of certain tasks. The effort that is made to complete them is accordingly compulsory. This compulsory effort, imposed for the sake of pre-determined ends, is work. The nineteenth century consequently deified work. It should be observed that such work consists in an unqualified effort, lacking any sort of prestige in its own nature, which derives its whole dignity from the necessity it serves. For this reason it has a homogeneous and purely quantitative character, which allows of its measurement by hours and its remuneration on a mathematically fixed scale.

Work is balanced by another kind of effort which does not arise from any kind of imposition, but is a perfectly free and hearty impulse of vital potency: this is sport.

If the final aim of the task which gives sense and value to effort is to be found in work, the spontaneous effort which dignifies the result is to be found in sport. The effort is a lavish one, which expends itself prodigally, without hope of recompense, as though it were an overflow of internal energy. Hence the quality of an effort made in the interests of sport is always of the finest. It cannot be subjected to the single standard of weight and measurement that regulates the ordinary remuneration of work. Tasks that are valuable are only completed through the mediation of this anti-economic type of effort: scientific and artistic creation, political and moral heroism, religious sanctity, are the sublime results of "sporting" efforts. But it should be noted that the progress to such results is not predetermined. No one has ever discovered a physical law simply by intending to do so: the discovery may more accurately be said to come to light in the guise of an unexpected windfall, a by-product of the worker's congenial and disinterested preoccupation with the phenomena of nature.

A life, then, which finds the exercise of its own powers more interesting and valuable than the prosecution of those aims which the taste of yesteryear garnished with so unique a prestige will give to its efforts the cheerful, hearty and even slightly waggish air that is peculiar to sport. It will diminish as far as possible the morose expression of the worker who alleges the justification of his toil in pathetic reflections on the duty of man and the sacred labour of culture. It will create its splendours as if in jest, and will not endow them with any great importance. The poet will manage his art with his toes, like a good footballer. The face of the nineteenth century bears throughout its extent the grim signs of a day of toil. Our present youth seems disposed to give life the careless aspect of a day of merrymaking.

It would not be difficult to point to signs of a similar type of variation in the political world. The most conspicuous feature of European politics during the last few years has been its depression. There is less political business done than there was in 1900; what little there is we carry on with less spirit and less industry. No one anticipates obtaining any satisfaction from it, and we are beginning to think our ancestors rather childish for letting themselves get killed at barricades for the sake of this or that formula of constitutional law. Or we might say, with more point, that the only admiration with which those frantic scenes now inspire us is directed to the gallant impulse that induced them thus to throw away their lives. Their motives, however, we can only regard as flimsy. Liberty is a conception that bristles with difficulties, and its value is nothing if not equivocal; on the other hand, heroism, that sublime "sporting" attitude, through which a man projects his life beyond its normal boundaries, possesses a vital grace which can never grow old. The public history of the last hundred and fifty years began with the oath taken in the tennis court.* One remembers the pictures painted of that glorious spectacle and the expressions of lofty earnestness with which the delegates performed their illustrious feat of declaration. The very act—an oath—reveals the fact that politics was then being given a religious importance. Everyone can now perceive the distance at which our own age stands from this mode of thought. Nevertheless, I repeat that it is not the case that political principles have lost their value and significance. Liberty still seems

* Of the Tuileries in Paris. The oath was taken by the Constituent Assembly of 1789. (Translator's note.)

an excellent conception to us, but it is no more than a plan, a formula or an instrument for living. To subordinate the latter to the former and deify the political idea of liberty is an idolatry.

The values of culture have not perished; but they do not now occupy the same rank as formerly. As soon as any new element is introduced into a perspective its whole hierarchy is recast. So, in the spontaneous system of valuations which the new race has brought into the world, which is, in fact, the new race itself, there has appeared a new value—vitality—and the mere fact of its presence is diminishing the rest. The epoch anterior to our own gave itself up in an exclusive and one-sided manner to the assessment of culture, and forgot all about life. The moment life is conceived as an independent value, subsisting apart from its contents, science, art and politics, though they may retain their original value, become less valuable in relation to the total perspective of the inward eye of mankind.

CHAPTER X

THE DOCTRINE OF THE POINT OF VIEW

TO oppose life to culture and demand for the former the full exercise of its rights in the face of the latter is not to make a profession of anticultural faith. If the foregoing observations have been interpreted in this way, the reader is facing in exactly the opposite direction to that required. The values of culture remain intact; all that is denied is their exclusive character. For centuries we have gone on talking exclusively of the need that life has of culture. Without in the slightest degree depriving this need of any of its cogency, I wish to maintain here and now that culture has no less need of life. Both powers—the immanent biological power and the transcendent power of culture—remain, when so considered, face to face on equal terms, neither being subordinated to the other. The mutual respect thus subsisting between the two permits the problem of their relations to be clearly defined and a more equitable and durable synthesis to be prepared.

Let us now recall the opening considerations of this discourse. Modern tradition presents us with a choice between two opposed methods of dealing with the antinomy between life and culture. One of them—rationalism—in its design to preserve culture denies all significance to life. The other—relativism—attempts the inverse operation: it gets rid of the objective value of culture altogether in order to leave room for life. Neither of these solutions, which appeared sufficient to the generations of the past, finds an echo in our own sensibility. Neither of them can live without being blind to the other's existence. Our own age, not being a prey to such obfuscations, and seeing with perfect clarity the significance of both contending powers, cannot bring itself either to accept the idea that truth, justice and beauty do not exist, or to forget that their existence requires the support of vitality.

Let us make this point clearer by concentrating upon that element in culture which is the easiest to define, viz., knowledge.

Knowledge is the acquisition of truths, and in acquiring truths we become acquainted with the transcendental or trans-subjective universe of reality. Truths are eternal, unique and invariable. How, then, can there be, in the knower, any process by which they can be identified? The reply of rationalism is narrow and arbitrary: knowledge is only possible if reality can penetrate it without the least disturbance of its own fabric. The knower, therefore, must be a transparent medium, lacking any sort of special quality or characteristic colour: he must be the same yesterday as to-day or to-morrow: he must therefore be ultra-vital and extra-historical. Life has essential characters of its own, it changes and develops: in a word, it is history.

The reply of relativity is equally narrow and arbitrary. Knowledge is impossible; there is no such thing as transcendent reality, for the reason that every real knower resembles an arena that has its own special formation. Reality would have to alter its own fabric in order to enter such an arena, and the particular alteration made would in each case be falsely construed as reality.

It is interesting to notice how in recent times, without any mutual collaboration or premeditation, psychology, biology and the theory of knowledge have each, in their survey of the facts which form the basis of both rationalist and relativist views, been obliged to make certain corrections, and are now unanimous in formulating the problem in a new way.

The knower is not a transparent medium, a pure Ego, possessed of fixed identity and an invariable nature, nor does his reception of reality result in disturbances of fabric in the latter. The facts impose a third view of the process of knowledge, which is a perfect synthesis of the other two. When a sieve or a net is placed in a current of liquid it allows certain things to permeate it and keeps others out; it might be said to make a choice, but assuredly not to alter the forms of things. This is the function of the knower, of the living being face to face with the cosmic reality of his environment. He does not allow himself, without more ado, to be permeated by reality, as would the imaginary rational entity created by rationalist definitions. Nor does he invent an illusory reality. His function is clearly selective. From the infinite number of elements which integrate reality the individual or receiving apparatus admits a certain proportion, whose form and substance coincide with the meshes of his sensitised net. The rest, whether phenomena, facts or truths, remain beyond him. He knows nothing of them and does not perceive them.

An elementary and purely physiological instance of this process may be found in the mechanism of sight and hearing. The ocular and auditive structures of the human race admit wave vibrations between fixed minimum and maximum velocities. Such colours and sounds as remain outside the two limiting points are unknown to humanity. In a similar way man's vital framework has a certain influence upon his reception of reality; but this does not mean that this influence or intervention involves alteration of the fabric of reality. A whole repertory, and a fairly large one, of perfectly real colours and sounds reaches his consciousness, and he is unquestionably aware of them.

The same process as operates in the case of colours and sounds applies also to truths. The psychic structure of each individual plays the part of a receptive organ in possession of a determinate form which admits the comprehension of certain truths and is condemned to an obstinate blindness to others. Similarly, all peoples and all epochs have their typical souls, that is to say, their nets, provided with meshes of definite sizes and shapes which enable them to achieve a strict affinity with some truths and to be incorrigibly inept for the assimilation of others. This means that all epochs and all peoples have been able to enjoy the measure of truth which suits them, and there is no sense in any people or epoch setting up in opposition to the rest, as if their particular share of truth were the respository of the whole of it. All have their fixed position in the historical series; none can legitimately aim at abandoning their posts, for such an act would be the equivalent of converting the agent into an abstract entity, and this would involve a total renunciation of existence.

Two men may look, from different view-points, at the same landscape. Yet they do not see the same thing. Their different situations make the landscape assume two distinct types of organic structure in their eyes. The part which, in the one case, occupies the foreground, and is thrown into high relief in all its details, is, in the other case, the background, and remains obscure and vague in its appearance. Further, inasmuch as things which are put one behind the other are either wholly or partially concealed, each of the two spectators will perceive portions of the landscape which elude the attention of the other. Would there be any sense in either declaring the other's view of the landscape false? Evidently not; the one is as real as the other. But it would be just as senseless if, when our spectators found that their views of the landscape did not agree, they concluded that both views were illusory. Such a conclusion would involve belief in the existence of a third landscape, an authentic one, not subject to the same conditions as the other two. Well, an archetypal landscape of this kind does not and cannot exist. Cosmic reality is such that it can only be seen in a single definite perspective. Perspective is one of the component parts of reality. Far from being a disturbance of its fabric, it is its organising element. A reality which remained the same from whatever point of view it was observed would be a ridiculous conception.

The case of corporeal vision applies equally to all our other faculties. All knowledge is knowledge from a definite point of view. Spinoza's *species aeternitatis*, or ubiquitous and absolute point of view, has no existence on its own account: it is a fictitious and abstract point of view. We have no doubt of its utility as an instrument for the fulfilment of certain requirements of knowledge, but it is essential to remember that reality cannot be perceived from such a standpoint. The abstract point of view deals only in abstractions.

This way of thinking leads to a radical reform in philosophy, and also, which is more important, to a reform in our sensuous reaction to the cosmos.

The individuality of every real subjective entity was the insurmountable obstacle encountered by recent intellectual tradition in its attempt to make knowledge justify its claim to be able to enter into possession of truth. Two different subjective entities, it was supposed, would acquire the knowledge of two divergent types of

DOCTRINE OF THE POINT OF VIEW

truth. We can now see that the divergence between the worlds of two subjective entities does not involve the falsity of one of them. On the contrary, precisely because what each one sees is a reality, not a fiction, its aspect must be distinct from what the other perceives. The divergence is not a contradiction, but a complement. If the universe had presented an identical appearance to the eyes of a Greek of Socrates' time and to those of a Yankee we should have to suppose that true reality, independent of subjective entities, does not reside in the universe. For the fact that it looked the same to two men placed at such diverse standpoints as those of Athens in the fifth century B.C. and New York in the twentieth A.D. would indicate that there was no question of any objective reality at all, but rather of a mere image which happened to occur, with identical features, in the minds of the two persons concerned.

Every life is a point of view directed upon the universe. Strictly speaking, what one life sees no other can. Every individual, whether person, nation or epoch, is an organ, for which there can be no substitute, constructed for the apprehension of truth. This is how the latter, which is in itself of a nature alien from historical variation, acquires a vital dimension. Without the development, the perpetual change and the inexhaustible series of adventures which constitute life, the universe, or absolutely valid truth, would remain unknown.

The persistent error that has hitherto been made is the supposition that reality possesses in itself, independently of the point of view from which it is observed, a physiognomy of its own. Such a theory clearly implies that no view of reality relative to any one particular standpoint would coincide with its absolute aspect, and consequently all such views would be false. But reality happens to be, like a landscape, possessed of an infinite number of perspectives, all equally veracious and authentic. The sole false perspective is that which claims to be the only one there is. In other words, that which is false is utopia, non-localised truth, which "cannot be seen from any particular place." The utopian (and such is essentially the character of the rationalist) goes further astray than anyone, since he is the spectator who loses confidence in his own point of view and deserts his post.*

Up to the present time philosophy has remained consistently utopian. Consequently, each successive system claimed to be valid for all ages and all types of mankind. Isolated beyond vital, historical and "perspectivist" dimension, it indulged from time to time in various unconvincing gestures of definition. On the other hand, the doctrine of the point of view requires a system to contain a properly articulated declaration of the vital perspective responsible for it, thus permitting its own articulation to be linked up with those of other systems, whether future or exotic. Pure reason must now give place to a vital type of reason in which its pure form may become localised and acquire mobility and power of self-transformation.

When we look to-day at the philosophies of the past, including those of the last century, we observe in them certain traces of "primitivism." I use the word in the strict sense in which it is applied to the painters of the Quattrocento. Why do we call them "primitives"? In what does their "primitive" quality consist? In their ingenuousness, in their candour, we say. But what is the reason for their candour and their ingenuousness,

* From the year 1913 onwards I have been expounding, in my university lectures, this doctrine of "perspectivity," which is briefly and arbitrarily formulated in *El Espectador*, I (1916). For the impressive confirmation of this theory in the work of Einstein see page 135.

what is the essence of these states of mind? Undoubtedly, self-forgetfulness. The "primitive" painter depicts the world from his point of view, that is, in obedience to ideas, valuations and sentiments which are peculiar to him; but he believes that he paints it as it is. For the same reason he forgets to introduce his own personality into his work; he offers us the work as if it had made itself, without the intervention of any particular agent; it is fixed at a definite position in space and at a definite moment in time. We naturally see in his picture the reflection of his own individuality, and thus do not see it as he did, since he took no account of himself as a person and believed himself to be the anonymous pupil of an eye spontaneously opening upon the universe. This habit of not taking any account of the self is the magic source of ingenuousness.

But the pleasure we derive from candour both includes and takes for granted a certain degree of disdain for the candid person. It is a benevolent enough disparagement. We enjoy the "primitive" painter as we enjoy the soul of a child, precisely because we are conscious of our own superiority. Our vision of the world is much ampler, more complex and more full of reservations, cross-roads and pitfalls. When we move in our vital circuit we are conscious of it as of something unlimited, uncontrollable, dangerous and difficult. On the other hand, when we approach the universe of the child or the "primitive" painter we perceive it as a tiny circle, quite shut in on all sides and quite manageable, furnished with a much smaller supply of objects and disguises. The imaginary life we lead during the period of this contemplation is for us a playful relaxation in which we can momentarily dispense with our own uneasy and troubled existence. The peculiar fascination of candour, then, is to be referred to the delight taken by the strong in the fragility of the weak.

The philosophies of the past have an attraction of a similar kind for us. Their clear and simple schematic pattern, their ingenuous illusion of being discoveries of truth in its entirety, the confidence with which they rely on formulæ which they imagine incontrovertible, convey the impression of a closed circle, defined and definitive, where there are no more problems to solve and everything is satisfactorily determined. There is nothing more pleasant than to spend a few hours in such clear and mild atmospheres. But when we return to our own thoughts and again react to the universe through our own particular sensibility we perceive that the world defined by the philosophies we have been examining was not really the world, but simply the horizon of the philosophers responsible. What they interpreted as the limit of the universe, beyond which there was nothing, was only the curve that closed the landscape their particular perspective afforded them. Every philosophy, which desires to eradicate this inveterate "primitivism," this persistent utopia, from its system, must correct the mistake I have referred to and avoid the eventuality of a malleable and expansible horizon hardening into a world.

Now, the reduction of the world to a horizon, or its conversion into one, does not lessen the quantity of reality in it to the smallest degree: the process simply puts it into relation with the living observer, whose world it is, endows it with a vital dimension and localises it in the current of life which flows from species to species, from people to people, from generation to generation and from individual to individual, gradually possessing itself of more and more universal reality.

Accordingly, the peculiar property of every living being, the individual difference, far from impeding the capture of truth, is precisely the organ by which the specially corresponding portion of reality is perceived. So that each individual, each generation or each epoch may be considered as an apparatus, for which there can be no substitute, directed to the acquisition of knowledge. Integral truth is only obtained by linking up what I see with what my neighbour sees, and so on successively. Each individual is an essential point of view in the chain. By setting everyone's fragmentary visions side-by-side it would be possible to achieve a complete panorama of absolute and universally valid truth. Now, this sum of individual perspectives, this knowledge of what each and all have seen and recognised, this omniscience, this true "absolute reason," is the sublime faculty which used to be attributed to God. God is also a point of view: but not because he possesses a watch-tower beyond the confines of the human area from which he can behold universal reality directly, as if he were one of the old rationalists. God is not a rationalist. His point of view is that of each one of us: our partial truth is also truth to him. Our perspective is veracious and our reality authentic to that extent. The only point is that God, as the catechism says, is everywhere and therefore enjoys the use of every point of view, resuming and harmonising in his own unlimited vitality all our horizons. God is the symbol of the vital torrent through whose infinite nets the universe gradually passes, being thus continuously steeped in and consecrated by life, that is to say, seen, loved, hated, painfully endured and pleasurably enjoyed by life.

Malebranche used to maintain that if we know any truth at all, it is because we see phenomena through God's eyes or from God's point of view. To me the inverse seems more probable, viz., that God sees phenomena through the medium of mankind or that mankind is the visual organ of divinity.

It is therefore peculiarly incumbent upon us not to defraud the sublime requirement that depends upon our co-operation for its fufilment, and, planting ourselves firmly in the position we find allotted to us, to open our eyes wide to our environment with a profound faith in our own organism and vital nature, and accept the labour that destiny assigns us—the modern theme.

SUPPLEMENTARY

THE SUNSET OF REVOLUTION

"The Sabbath was made for man, not man for the Sabbath."—*S. Mark ii, 27.*

FOR the purpose of defining an epoch it is not enough to know what has been done in it; it is also necessary for us to know what it has not done, what was, in fact, impossible in it. This may seem a singular requisite; yet such is the condition upon which our thought proceeds. To define is to exclude and deny. The more reality what we define may possess, the more exclusions and denials we shall have to practise. Accordingly, the most profound definition of God, the supreme reality, is that given by the Indian Yajnavalkya: "Na iti, na iti." "Nothing of that kind, nothing of that kind." Nietzsche acutely observes that we are more influenced by what does not happen to us than by what does and, according to the Egyptian ritual of the dead, when the "double" abandons the corpse and has to perform its feat of self-definition before the judges of the world beyond the grave, it makes its confession contrariwise, that is to say, it enumerates the sins it has not committed. Similarly, when we declare that one of our acquaintances is an excellent person, do we mean anything except that he will not rob or kill us, and that if he does covet his neighbour's wife no one will be very much concerned about it?

The positive character with which we thus invest negation is nevertheless not simply a necessity inflicted by the peculiar temper of our intelligence. There is, at any rate in the case of living beings, a real vigour of negation which corresponds to the negative concept. If the Romans did not invent the motor-car, that was no mere accident. One of the ingredients that play a part in Roman history is the incapacity of the Latin race in matters of technical detail. This was one of the most active factors in the decadence of the antique world.

An epoch is a repertory of positive and negative tendencies; it is a system of subtleties and perspicacities united to a system replete with blindness and dullness. There is not only the taste for certain things, but also the determination to have distaste for others. At the beginning of a new age the first thing we notice is the magical presence of these negative propensities, which initiate the elimination of the fauna and flora of the anterior epoch: so, in the flight of the swallows and in the fall of the leaves we first become aware of the presence of autumn.

In this sense there is no better qualification of the age now dawning upon our ancient continent than the recognition that in Europe revolutions are things of the past. Such recognition implies not only that they no longer exist in fact, but also that they can never exist again.

Perhaps the full significance comprised in this prognostication does not appear obvious at once for the reason that the current notion of revolution is a very vague one. Not long ago an excellent friend of mine, of Uruguayan nationality, assured me, with ill-concealed pride, that in less than a century his country had undergone forty revolutions. Evidently my friend was exaggerating. Educated, like myself and a good number of my readers, in an uncritical worship of the idea of revolution, he patriotically desired to adorn his national history with the greatest possible number of concrete instances. To this end, following a common custom, he called revolutionary every collective movement in which violence is employed against established power. But history cannot be content with such rough and ready notions. It requires more exact instruments and more sharply outlined concepts for its purpose of sound orientation in the forest of human occurrences. Not every violent measure against public power is revolution. It is not, for example, revolution when one part of society rebels against the governing class and violently substitutes others for them. The convulsions of the South American peoples are almost always of this type. If there is a very earnest desire to retain the title of revolution for them, we should not dream of inspiring a further example in order to thwart the desire in question: but we shall have to look for another name to denominate another class of processes of an essentially distinct type, to which belong the English revolution of the seventeenth century, the four French ones of the eighteenth and nineteenth, and in general all the public life of Europe between 1750 and 1900, which Auguste Comte had already, by 1830, proved to be "essentially revolutionary" in character and derivation. The same motives which induce people to think that there will be no more revolutions in Europe oblige them to believe that there have not yet been any in South America.

The least essential feature of true revolution is violence. It is not inconceivable, though it is hardly likely, that a revolution might run its whole course without a drop of blood being shed. Revolutions are not constituted by barricades, but by states of mind. Such states of mind do not occur in all ages; like fruits, they have their seasons. It is a remarkable fact that in all the great historical cycles of which we have sufficient knowledge—the Greek, Roman and European worlds, for instance—

a point is reached at which not one revolution but a whole revolutionary era begins, which lasts two or three centuries before it finally disappears for good.

It shows a complete lack of historical perception to consider the risings of peasants and serfs in the Middle Ages as events foreshadowing modern revolutions. There is no real connection between the two types of phenomena. When the medieval man rebels it is against the abuses indulged in by his lords. The modern revolutionary, on the other hand, does not rebel against abuses but against usage or custom. Up to a short time ago histories of the French revolution began by representing the years round about 1780 as a time of misery and social depression, with affliction rife in the lower classes and tyranny in the upper. In their ignorance of the specific structure of revolutionary eras people believed that the catastrophe could only be explained as a movement of protest against an antecedent oppression. It has now long been recognised that in the stage previous to the general rising the French nation enjoyed greater wealth and more even-handed justice than in the time of Louis XIV. It has been declared a hundred times that the revolution was formed in men's minds before it began in the streets. If a sound analysis had been made of what is implied by this expression, the physiology of revolution would have been discovered.

All revolutions, in effect, if they are true revolutions, presuppose a peculiar and unmistakable disposition of mind. To understand it properly one must turn to study the development of the great historical organisms which have completed their full cycle. We then find that in every one of those great composite movements mankind has passed through three distinct spiritual situations, or, in other words, that the life of the human psyche has gravitated successively towards three diverse centres.*

The psyche passes from a traditional state of mind to a rationalist and from the latter to a mystical *régime*. These are, so to speak, the three different forms of psychic mechanism, the three distinct ways in which the mental apparatus of mankind pursues its function.

During the centuries in which some great historical conglomeration, such as Greece, Rome or our own continent of Europe, is in process of formation and organisation, what *régime* directs the spirit of its members? The answer given by facts is a most surprising one. It is when a people is young and in course of development that it is chiefly influenced by the past. At the first glance the contrary would appear to be the more natural state of affairs: one would suppose that an ancient people, with a long past behind them, would be most thoroughly subject to the claims of bygone days. This, however, is not the case. The decrepit nation is not in the slightest degree influenced by the past; on the other hand, in an adolescent population everything is done with an eye to the past. And it is not a short past that is envisaged, but one so long, and with so vague and remote a horizon, that no one has ever seen, or remembers, its commencement. It is, in brief, immemorial.

* Strictly speaking, I ought to distinguish many more modifications of the human psyche throughout a complete historical cycle; and if I mention only three the trinity must not be taken to possess any cabalistic virtue. It signifies merely that by concentrating on three extreme forms of psychic evolution we obtain sufficient points of reference to throw light on the vast historical phenomenon with which we are now concerned. If we were dealing with a phenomenon of smaller proportions we should have to draw nearer to the historical area and the three main headings would then be subdivided into many more. Concepts which coincide with reality when the latter is contemplated at a certain distance have to be replaced by others when the distance is shorter, and *vice versa*. Thought proceeds under the guidance of a law of perspective corresponding to that operative in vision.

The psychology of peoples dominated by ancestral ideas and arrested, through one kind or another of historical malnutrition, in a permanently infantile stage of development is a curious study. One of the most primitive peoples in existence is the aboriginal Australian. If we investigate the way in which the intellectual activity of this people functions, we find that on being confronted with any sort of problem—for example, a phenomenon of nature—the Australian does not look for an explanation which is enough of itself to satisfy intelligence. In his mentality, to account for a fact such, for instance, as the existence of three rocks standing together on a plain, is to recall a mythological story which he has heard ever since he was a child, and according to which in antiquity, or, as the Australians say, in *alcheringa*, three men, who were once kangaroos, were changed into the stones in question. This explanation satisfies his mind precisely because it is not a reason or a thought which can be verified. Its validity consists in the fact that the individual intelligence creates it for itself, either as an original statement, or by repeating the ratiocination and observations which integrate it. The strength of reason is born of the conviction that it produces in the individual. Now, the Australian does not experience what we call individuality or, if so, he experiences it in the form and to the extent that a child does when it is left alone, abandoned by the family group. The primitive man only perceives the singularity of his person as solitude or disruption. The concept of individuality and everything based upon it only produces terror in him: it is a synonym, for him, of debility and insufficiency. Solidity and security are to be found only in the communal condition, whose existence is anterior to that of any individual: for the latter finds it ready-made for him as soon as he awakes to life. As the old men of the tribe had been equally conscious of it, it is considered to be of immemorial origin. It does each man's thinking for him by means of its treasure of myth and legend, transmitted by tradition; it creates his legal and social codes, his rites, dances and gestures. The Australian believes in the mythological explanation precisely because he has not invented it, precisely because he does not possess a sound reasoning faculty. The reaction of his intellect to the events of life does not consist in the immediate expression of a spontaneous thought of his own, but in reiterating a pre-existent and accepted formula. For these people thought, desire and feeling connote mere circulation through ready-made psychological channels, repetition of a hackneyed repertory of mental attitudes. The spontaneous, in this mode of existence, is fervent submission and adaption to accepted type, to the tradition in which the individual lives submerged, and which is, for him, immutable reality.

This is the traditionalist state of mind which has been operative in our own Middle Ages, and which directed the course of Greek history up to the seventh and Roman to the third century B.C. The content of these epochs is naturally much richer, more complex and more delicate than that of the mind of a savage; but the type of psychic mechanism and its method of functioning is the same. The individual invariably adapts his reactions to a communal repertory which he has received by transmission from a venerated past. The medieval man, when he has to decide upon a course of action, puts himself into relation with what his "fathers" did. The situation is identical, in this respect, with that prevailing in the mind of the child. The child, too, believes more in what it hears from its parents than in its own judgments. When an event is described in the presence of children they generally direct an interrogative look at their parents, as if to ask them whether the narrative is to be believed, whether it is "true" or a "story." The mind of the child, too, never gravitates to the centre of its own individuality: it clings to its progenitors in the same way as the medieval mind clings to the "usage and custom of our fathers." In no system of jurisprudence does customary law, or immemorial usage, possess such weight as it does in the systems which arise during historical formations and consolidations. The simple fact of antiquity is converted into a legal sanction. The foundation of jurisprudence is neither justice nor equity, but the irrational, by which I mean the purely material, fact of prolonged existence.

In the political world the traditionalist mind will be found living in respectful concurrence with what is already established which, precisely because it is established, possesses an invulnerable prestige: it is what we find ready-made for us when we are born; it is what our fathers did. When a new requirement presents itself it does not occur to anyone to reform the structure of established fact; what is done is to make room in the latter for the new fact and give it a permanent place in the immemorial body of tradition.

It is in the epochs characterized by the traditionalist mind that nations organize themselves. For this reason such periods are followed by an age of maturity, which is, in a certain sense, the hour of historical culmination. The body of the nation has reached its perfect development: it enjoys the use of all its organs and has accumulated a vast treasure of energies together with potentialities of a high order. A time comes when all this wealth begins to be expended, and such stages of history then appear to us particularly healthy and brilliant. We are more forcibly aware of our neighbour's health when he begins to turn it to account externally in various outstanding exploits or, in other words, when he begins to lose it by expenditure. Such ages are the splendid centuries of vital dilapidation. The nation is no longer content with its internal life, and an epoch of expansion is initiated.

With such an epoch coincide the first clear symptoms of a new state of mind. The traditionalist mechanism of the mind is about to be replaced by another mechanism of an opposite type—the rationalist.

We, too, in the present age are affected by traditionalism; but we must avoid confusing this type with what I have previously called traditionalism in this essay. Contemporary traditionalism is no more than a philosophic and political theory. The traditionalism of which I have spoken, on the contrary, is a reality: it is the real mechanism responsible for the functioning of men's minds during certain epochs.

So long as the empire of tradition lasts, each unit of mankind remains embedded in the close corporation of communal existence. He does nothing on his own account, apart from the social group. He is not the protagonist of his own acts; his personality is not his own, distinct from others; an identical mind is reproduced in each unit with the same thoughts, memories, desires and emotions. Hence, in traditionalist centuries figures of outstanding personal physiognomy are not, as a rule, to be found. All the members of the social body are more or less the same. The only important differences are those of position, rank, employment or class.

However, within this communal mind, whose texture is that of tradition, and which has its seat in each unit of the group, a small central nucleus begins, after a time, to form: this is the sentiment of individuality. It originates in a tendency antagonistic to that which the traditional mind has been moulding. The supposition, that the consciousness of private individuality was a primary notion and, so to speak, aboriginal in man, was quite erroneous. It used to be asserted that human beings are originally aware of themselves as individuals, and that the next step is to seek out other human beings with the object of associating with them. The truth is just the opposite. The subjective personality begins by feeling himself to be an element of a group, and it is only later that he proceeds to separate from it and achieve little by little the consciousness of his singularity. The "we" comes first, and then the "I." The latter is therefore endowed from its birth with the secondary character of secession. I mean by this that man proceeds to discover his individuality in proportion to the development of his conscious hostility to communism and opposition to tradition. Individualism and anti-traditionalism are one and the same psychological force.

This nucleus of individuality, germinating within the traditionalist mind like the larva of an insect in the core of a fruit, gradually grows to the dimensions of a new demand, principle or imperative, confronting tradition. On this view the traditional method of reacting intellectually—I hardly care to call it thought—consists in recalling the repertory of beliefs received from the forefathers of the group. On the other hand, the individualist method turns its back on all such accepted beliefs, repudiating them just because they are accepted, and aims instead at producing some new thought which is to be valued on the grounds, only, of its own independent content. Such a thought, not proceeding out of immemorial communistic life, not to be referred to "our fathers," an ideation lacking lineage, genealogy and the prestige of hereditary emblems, is obliged to derive its parentage from its own works, to sustain itself by its own convincing efficacy, by its purely intellectual perfections. In a word, it must be a Reason.

The traditionalist mind functioned under the guidance of a single principle and possessed a unique centre of gravity which was, in fact, tradition. But henceforward two antagonistic forces operate in the mind of each unit, viz., tradition and reason. Little by little the latter will go on gaining ground from the former: this means that spiritual life has been converted into an internal struggle and has exchanged its unitarian status for dissociation into two mutually inimical tendencies.

While the primitive mind accepts, as soon as it is born, the world which it finds already provided for it, the birth of individuality involves at once a negation of that world. But the subjective personality, in repudiating the traditional, finds itself obliged to reconstruct the universe through its own resources, i.e., its reason.

It is easy to see that in consequence of this necessity the human spirit may succeed in developing its intellectual faculty to a point nothing short of marvellous. These rationalist periods are always the most glorious epochs of human thought. The irrational myth is put on the shelf, and in its place the scientific conception of the cosmos proceeds to the erection of its admirable edifices of theory. The specific enjoyment to be derived from ideas makes itself felt, and an amazing virtuosity in their invention and management is acquired.

Man ends by believing that he possesses a sort of divine faculty capable of revealing to him, once and for all, the ultimate essence of phenomena. This faculty must be independent of actual experience, whose constant variations might induce modifications in the revelation expected. Descartes called this faculty *raison* or *pure intellection*, and Kant, more accurately, pure reason.

"Pure reason" is not the ordinary exercise of the understanding, but its method of functioning absolutely. When Robinson Crusoe applies his intelligence to the resolution of the urgent problems that await him on his desert isle he does not employ pure reason. He imposes on his intellect the task of adjusting itself to circumambient reality, and its actual function is reduced to the effective combination of truncated portions of such reality. Pure reason is, on the contrary, the state of the understanding when left to its own devices: it then constructs, on its own foundations, a number of prodigious weapons, of a sublime accuracy and rigidity. Instead of seeking contact with phenomena it ignores such contact, and tries to ensure the most exclusive fidelity to its own internal laws. Mathematics is the typical product of pure reason. Its concepts are elucidated once and for all, and there is no risk of reality contradicting them at some future date, for reality is not their source. In mathematics nothing is uncertain and approximate. Everything is clear, for everything stands at its highest point of expression. Greatness is infinite greatness, and smallness is absolute smallness. The straight line is radically straight, and the curve curves unadulterated. Pure reason never passes beyond the circle of superlatives and absolutes. Indeed, that is the reason why it is called pure. It is incorruptible and uncompromising. When it defines a concept it endows it with perfect attributes. It can only think in terms of the utmost limit, i.e., radically. As its operations are entirely self-reliant, it can give its creations the maximum polish without going to very much expense. In the same way, in the realm of political and social questions, it is in the habit of believing that it has discovered a civil constitution or a code which is perfect and definitive, and which alone deserves the names. This pure use of the intellect, this thinking *more geometrico* is generally called rationalism. Perhaps it would be more enlightening to call it radicalism.

Everyone is unanimous in recognising that revolutions are not in essence anything but political radicalism. But perhaps it is not everyone who perceives the true sense of this formula. Political radicalism is not an original attitude: it is rather a consequence. It is not *radical* in politics because it is radical in *politics*, but because it is already radical in thought. This distinction, though it may have a frivolously super-subtle air, is decisive for the comprehension of the historical phenomenon which is properly styled revolutionary. The scenes which such phenomena invariably exhibit are signalised by such moving evidences of passion that we feel inclined to seek the origin of revolution in emotion. Some people will see the motive power of the impressive event in the explosion of a certain type of civic heroism. But Napoleon used to say: "Vanity made the revolution: liberty was only the pretext." I do not deny that both of these passions may be ingredients of revolution. But in all the great historical epochs there have been plenty of instances of heroism and vanity which do not necessarily lead to catastrophic outbreaks. For revolution to result from the operation of these two affective forces they must function in a spirit saturated with faith in pure reason.

This consideration enables us to account for the fact that in every great historical cycle a moment arrives when the revolutionary mechanism suddenly begins to act with uncontrollable violence. In Greece as in Rome, in England as on the continent of Europe, intelligence, in the pursuit of its normal development, reaches a stage at which it discovers its power of constructing, with means exclusively its own, theoretical edifices on a large scale and perfect in form. It had previously existed entirely upon the observations of the senses, which are for ever in a state of fluctuation, *fluctuans fides sensuum*, as Descartes, the father of modern rationalism, used to say, or upon the sentimentally interpreted prestige of political and religious tradition. But there now suddenly appears one of those ideological specimens of an architecture constructed by pure reason, such as the philosophic systems of the Greeks of the seventh and sixth centuries, the mechanics of Kepler, Galileo and Descartes, or the *Natural Law* of the seventeenth and eighteenth centuries. The transparency, precision, rigidity and integrity in systematisation of these compact spheres of ideas, manufactured *more geometrico*, are incomparable. From the intellectual point of view nothing more estimable can be imagined. It should be noted that the qualities I have enunciated are specifically intellectual; they might be called the professional virtues of the intelligence. It is, of course, true that there are many other values and attractive qualities in the universe which have nothing to do with the understanding, e.g., fidelity, honour, mystic fervour, solidarity with the past, authoritative power. But when the great rational creations arise men are already a little tired of such values. The new qualities, of an intellectual category, make an ardent and exclusive appeal to the human spirit. The result is a strange disdain for realities: men turn their backs to the latter and become the impassioned slaves of ideas as such. The perfection of the geometrical form of the idea intoxicates its devotees to the point of forgetting that, by definition, the business of the idea is to coincide with the reality of which it is the expression in the medium of thought.

The next step is the total inversion of spontaneous perspective. Ideas have so far been employed simply as instruments in the service of vital necessities. But now life is to take up the service of ideas. This radical reshuffling of the relations between life and idea is the true essence of the revolutionary spirit.

The subversive movements of the burgesses and peasants of the Middle Ages did not aim at the transformation of the political and social *régime* of the period: quite the reverse: they either limited themselves to accomplishing the reform of some abuse, or their object was the attainment of certain particular benefits or privileges within the framework of the established *régime;* they thus signified their approval of its general configuration. No one moderately well informed would venture to-day to compare the guilds and corporations of the thirteenth and fourteenth centuries with modern democracies. The latter have, it is true, appropriated much of the juridical technique that the guilds and corporations elaborated; but the spirits of the ancient and modern institutions are utterly different. It was with good reason that the city constitutions were called "charters" or "privileges" in Spain. The endeavour they symbolised was, precisely, the attempt to adjust the established *régime* to new necessities and desires, the idea of jurisprudence to life. The charter is a privilege, that is to say, it is a legally constituted vent for the new energy in the system of traditional powers. The point is that such energy, instead of transforming the system, is assimilated to it and implanted in its structure. The system, on its side, yields to and admits the newly introduced reality.

The political principles of the medieval burgess involved no more than the establishment, in opposition to the privileges of the nobility, of further privileges of similar type. The city guilds and the various corporations prided themselves on the possession of an even narrower, more suspicious and more egotistical spirit than that of the feudal lords. The highest authority on the life of the citizen in the Middle Ages—the Belgian, Henri Pirenne—notes that the corporations, in their most democratic epoch, practised an exclusiveness in their political life of an almost incredible character, and showed less hospitality to strangers and newcomers than had ever existed before. So much was this the case that "while the neighbouring rural communities increase in density the statistics of the burgess population within the city walls show no increase whatever." The strange phenomenon of a sparse urban population during these centuries is accordingly due to the resistance offered by the towns to the influx of fresh competitors for their liberties. "Far from trying to extend their legal code and institutions so as to include any considerable portion of the peasantry, the towns were more jealous in guarding their monopolies the more the popular *régime* achieved consolidation and development within their bounds. They endeavoured, moreover, to impose an extremely burdensome hegemony on the people of the free rural districts, treated them like subjects and, when the opportunity arose, violently compelled them to sacrifice themselves for the benefit of their dictators." "In short, then, we may conclude that the urban democracies of the Middle Ages were not and could not be anything but democracies comprising a privileged membership." Now, democracy in the modern sense and privilege are the most complete contradiction that can be imagined. "It is not," pursues Pirenne, "that the theory of democratic government was unknown to the Middle Ages. The philosophers of the time formulated it clearly, in imitation of the ancient political writers. In Liége, in the midst of civil dissensions, the good canon Jean Hocsem examines quite seriously the respective merits of aristocracy, oligarchy and democracy, and finally gives his verdict in favour of the latter. Moreover, it is sufficiently well-known that more than one scholastic philosopher has formally recognised the sovereignty of the people and their right to dispose of political power. But these theories did not exercise the least influence upon the contemporary *bourgeoisie*. Their influence can, no doubt, be traced, during the fourteenth century, in certain political pamphlets and in a few literary works; it is, however, perfectly certain that they had not, at any rate in the Low Countries, the smallest influence on the Commune."*

The idea that some "radicals" in Spain have had of connecting their own democratic politics with the rise of civic communities in the Middle Ages merely reveals the ignorance of history which is a permanent attribute, like some innate vice, of radicalism.

Modern democracy is not to be ascribed directly to any ancient democracy, neither the medieval nor the Greek nor the Roman. The only legacies of the classical democracies to our own age have been a misrepresented terminology, the general cast of their features, and their rhetoric.† The procedure of the Middle Ages was to amend the existing *régime*. That of our own era, on the other hand, has been to organise revolutions; that is to say, instead of adapting *régime* to social reality we have made attempts to adapt the latter to the scheme of an ideal.

---

* Henri Pirenne: *Les Anciennes Democraties des Pays-Bas*, pp. 133, 197, 199, 200.

† The analysis of the differences between our own democracies and those of other times, as well as the study of their genesis, must await a future occasion. The whole subject is a prey to the most confused notions. I have asked many eminent radicals what they understand by democracy and liberalism, but have never obtained any replies that did not, by their vagueness, discount acceptance. Yet the two concepts are perfectly clear, though it is true that their evident genealogy is the very last that a practical democrat would suspect.

When the feudal lords, in their hunting expeditions, gallop over the farmer's crops and destroy them, the farmer feels a natural irritation and is anxious to retaliate, or at any rate to avoid such a misfortune in the future. But it does not occur to him that in order to prevent the repetition of so concrete an injury to his property or person it may be necessary to bring about a radical transformation of the entire organisation of society. In our own time, on the contrary, the deep resentment of the oppressed citizen is directed not so much against the oppressor himself as against the whole architecture of a universe in which such oppression is possible. For this reason I maintain that while the medieval man is irritated by abuses—of a *régime*—the modern man is irritated by usage, that is to say, by the *régime* itself.

The desire of the rationalist temperament is to mould the social body, at all costs, to the pentagraph of concepts framed by pure reason. In the view of the revolutionary the value of the law is pre-existent to its suitability to life. The good law is good by its own nature, like a pure idea. Accordingly, for the last century and a half European politics have been almost exclusively politics of ideas. A political philosophy concerned with realities and involving no anxiety for the triumph of an idea as such has hitherto seemed immoral. I do not at all mean by this that a political philosophy of private interest and ambition may not, in fact, have been practised surreptitiously. But the symptomatic feature of the matter is the fact that the political philosophy in question could not keep on its course and make its way without assuming the sanction of idealist colours and masking its true intentions.

Now, an idea framed without any other object than that of perfecting it as an idea, however much it may conflict with reality, is precisely what is called utopia. The geometrical triangle is utopia; nothing visible and tangible exists in which the definition of the triangle can find exact expression. Accordingly, utopianism is not an affection peculiar to a certain political doctrine, but the character appropriate to what pure reason elaborates. Rationalism, radicalism and the exercise of thought *more geometrico* are examples of utopianism. In science, perhaps, which is a contemplative function, utopianism may have a necessary and lasting mission to perform. The practice of politics, however, is a matter of realisation. How is it that the utopian spirit has not been found incompatible with politics?

The fact is that every revolution cherishes the entirely chimerical object of realising a more or less complete utopia. The plan inevitably fails. Its failure creates the twin and antithetical phenomenon of all revolutions, viz., counter-revolution. It would be interesting to prove the latter no less utopian than its antagonist and sister, even when less inspiring, warm-hearted and intelligent. Enthusiasm for pure reason will not admit defeat and returns to the charge. Another revolution breaks out, with yet another utopia, a modification of the first, inscribed upon its banners. There is a fresh failure and a fresh reaction; and so it goes on until the social conscience begins to suspect that the ill-success of these attempts is not due to the intrigues of their enemies, but to the contradictory elements inherent in the objects aimed at. Political ideas lose their glamour and attractive force. All that is facile and puerile in their schematic organisation begins to come to light. The utopian programme reveals its underlying formalism, its poverty and aridity in comparison with the delicious, abundant and splendid stream of life. The revolutionary era ends very simply, without phrases or gestures, in reabsorption by a new sensibility. To the political philosophy of ideas succeeds a political philosophy of concrete phenomena and men. We discover at last that life does not exist for the benefit of the idea, but that the idea, the institution, the rule exist for the benefit of life, or, as the Gospel has it, that "the Sabbath was made for man, not man for the Sabbath."

In particular—and this is a very important symptom—the whole business of politics comes to lose its character of urgency, disappears from the foreground of human interests and is permanently converted into a necessity of the same type as so many others, unavoidable, but not inspiring and not likely to be served with any degree of solemn and quasi-religious veneration. For it should be observed that in the revolutionary era political philosophy is found installed in the very centre of human preoccupations. There is no better apparatus for the registration of the hierarchy of our vital enthusiasms than death. The most important thing in our lives will always be that for which we are capable of dying. And the modern man has, in point of fact, risked his life at the barricades of revolution, thereby showing unmistakably that he expected politics to provide him with happiness. When the sunset of revolution commences this fervour of the previous generations appears to most people to be an evident aberration of the perspective of sentiment. Politics is not susceptible of exaltation to such high rank among hopes and devotions. The rationalist mind wrecked political science by expecting too much from it. When this thought begins to become general it puts an end to the era of revolution, to the political philosophy of ideas and the struggle for constitutional right.

The process has always been the same in Greece, in Rome, and in Europe. Laws are at first the effect of necessities, forces or dynamic combinations, but are soon converted into the expression of illusion and desire. Have juridical forms ever furnished men with the happiness they expected from them? Have the problems that originated them ever once been solved? Such are the suspicions now germinating at the roots of European consciousness and initiating a new type of spiritual mechanics which will replace the rationalist type as the latter supplanted that of the traditionalists. An anti-revolutionary epoch is beginning; but short-sighted people believe that a universal reaction is setting in. I am unaware of a single epoch of reaction throughout the whole era of history; there has never been such a thing. Reactions, like counter-revolutions, are casual and altogether transitory intervals, which derive their sustenance from vivid memories of the latest rebellion. Reaction is no more than a parasite of revolution. Such movements have already commenced in the southern periphery of Europe, and it is extremely probable that they will soon extend to the great nations of the centre and the north. But all that will be fugitive in character, little more than the noticeable oscillation that always precedes arrival at a new state of equilibrium. The revolutionary mind has never been succeeded in history by a reactionary mind, but rather—a more simple matter—by a disillusioned one. It is the inevitable psychological legacy of the splendid centuries of idealism and rationalism; those periods of organic dilapidation intoxicated with faith and self-assurance, those great topers of the beverages of utopia and illusion.

The physiognomy of the human mind in its traditionalist and revolutionary aspects, such as I have delineated it above, is undoubtedly in harmony with the development of European history from 1500 to our own day. The principal events of the latter centuries are too widely known for their concrete evidence to have failed to authenticate in the reader's mind the general scheme I have outlined of the configuration of the revolutionary spirit. But it is more interesting, it may even be called somewhat exciting, to observe that the same scheme is exactly reproduced in the other historical cycles of which we have any fairly definite knowledge. After this discovery the spiritual phenomenon of revolution acquires the character of a cosmic law of universal application, a stage through which every national body passes, and the transition from traditionalism to radicalism comes to resemble a biological rhythm pulsating irresistibly, so to speak, throughout history, after the manner of the rhythm of the seasons in vegetable life.

Let us, then, recall certain events in Greek and Roman history which fit with rare precision into the scheme I have described, and constitute its most adequate proof. This course will allow me, at the same time, to transcribe one or two paragraphs from great historians who, preoccupied exclusively with their own requirements and not, like myself, on the watch for historical generalisations, describe this or that moment in the life of Greece and Rome. If these authors, without troubling to look very closely at what they were doing, and without premeditation, have found themselves compelled to postulate behind the concrete case they are narrating the same mechanism of revolutionary spirit which I have defined as a universal stage in history, the coincidence will not be denied a demonstrative value of high rank.

In Greek and Roman history, up to some considerable time ago, an error was allowed to persist which is only now beginning to be corrected. Fundamentally, it was a belief that the hour of prosperity in Greece and in Rome coincides with the epoch which is the source of our abundant historical material. The whole of the earlier period was considered to have been a time of racial consolidation, prehistorical in the case of both nations. By an optical illusion very frequent in this field of investigation history confounds the non-existence of dates with the non-existence of events. A rectification of the error in question showed that the reality was very different from what had been supposed. The epochs concerning which a great deal of information begins to be accumulated are epochs in which historians already exist who undertake its preservation. Now, when historians begin to be found among a people it means that the people in question has already ceased to be young, that it is actually fully mature and may be taking its first steps to decadence. History, like the grape, is an autumnal delicacy.

The age at which the life of Greece and Rome becomes perfectly clear to us is already their September. The true history of the earlier period of these peoples, their youth and infancy, remains practically untouched. Accordingly, the face of the Greco-Roman image so ecstatically worshipped by the last few centuries was long past its prime; wrinkles had already installed upon it the geometrical designs which are the first indications of a cadaverous rigidity announcing the decline of life.

Mommsem was the first to rectify the perspective of Roman history. The great Eduard Meyer did the same, but to a more limited extent, with that of Greece. To the latter is due one of the most important and fertile innovations of historical thought. The division of universal history into ancient, middle and modern periods was a pentagraph dictated by convention and caprice and has, from the seventeenth century onwards, been hammered hard, so to speak, into the continuous body of history. Reconstructing Hellenic life, Meyer found that the Hellenes had passed through an epoch not unlike our own Middle Ages, and he ventured to speak of it as the Greek Middle Age. This discovery involved the transposition of the three ages of history to the historical cycle of each nation. Every people has its ancient, medieval and modern age. The practice of this view of history completely alters the significance of the traditional division into periods, and its three stages cease to be external, conventional or dialectical labels and assume a more real and, so to speak, biological significance. They are the infancy, youth and maturity of each people.*

The Greek Middle Age comes to an end in the seventh century. This is the first period concerning which we possess any copious and exact information. There is, however, no question here of the birth of a nation. On the contrary, we are invited to witness the protracted dissolution of a people's long past and its awakening to a new age. Meyer sums up the position as follows: "The foundations of the medieval political constitution are destroyed. The dominion of the nobles is no longer an adequate expression of the prevailing circumstances; the interests of the governing and governed classes no longer coincide. The antique pattern of life, of law and of communities founded upon consanguinity loses its significance and becomes an obstruction. Men no longer necessarily remain members of the circle in which they were born. Everyone is master of his own fate; the individual emancipates himself socially, spiritually and politically. If a man cannot make his fortune in his own country he goes to seek it among foreigners. Affairs involving currency and revenues—the economics of finance begin during this epoch—are considered immoral, and

* This idea of Meyer's largely inspired Spengler's suggestive work, *The Decline of the West*.

everyone becomes aware of their disastrous effects; but no one can be indifferent to them, and the most conservative nobleman takes good care not to despise his profits. *Chremata, chremata aner*—money, money makes the man— is the motto of the times; and it is very significant that we find it put into the mouth of a Spartan (Alcaeus, frag. 49) or of an Argive (Pindar, Isthmians 2). Between the nobles and the labourers come the new industrial and mercantile classes, with their attendant corps of artisans, petty traders and seamen, among whom are conspicuous such adventurers as Archilochus of Thasos, who seek their fortunes wherever they can, and bear the double burden of calamity and subjection to an alien power. The cities grow bigger, for the peasants migrate to them so as to secure an easier livelihood; foreigners, too, who had no luck in their own country or had to go into exile on account of party struggles, settle in the towns. All combine in attacks upon the *régime* of the aristocrats. The peasants aspire to freedom from the intolerable burden of economic oppression; the newly rich citizens to participation in administrative power; the descendants of the immigrants, who are sometimes more numerous than the longer established citizens, claim equality of treatment with the hereditary inhabitants. All these elements are united under the name of *demos*, as they were during the French Revolution under the name of *tiers état*. Like the latter, the Greek *demos* does not constitute a unity, either through its position or through its political and social aims; it is only the common opposition of such heterogeneous elements to the 'better sort' that maintains their alliance."*

There can be no more exact parallel with the composition of modern nations on the eve of the revolutionary

* Eduard Meyer, *Geschichte des Altertums*, Vol. II.

era. The more general distribution of money introduces capitalism. The rise of the latter is accompanied by that of imperialism. Soon afterwards comes the creation of great fleets. The wars of the mounted medieval nobles—I am now speaking of Greece—are replaced by others, not conducted on horseback or man to man. The *promachia*, or single combat, is succeeded by the great invention of the phalanx of hoplites, the body of infantry capable of tactical movements. At the same time the medieval system of dissociated units is brought to an end and all the Greeks begin to call themselves Hellenes. Under the unity of this name they come to feel their profound historical affinity.

Finally, it is in this age that the abrupt legislative changes in constitutions are introduced. Can the fact that these "invented" constitutions are always coupled with the name of some philosopher be due to mere chance? For it is, let us not forget, the century of the Seven Wise Men, and of the first Ionian and Dorian thinkers. Where there is radical alteration of laws and the establishment of new codes of conduct there is also to be noted, invariably, the manifest or covert presence of some "wise man." The Seven Wise Men are the seven great intellectual leaders of the epoch, the discoverers of reason or *logos* as opposed to *mythos* or tradition.

By a rare piece of good fortune our data enable us to witness, through documentary evidence, the first incarnation of the individualist and rational mentality in revolt against the mentality of tradition. The first thinker whose figure has come down to us in the light of complete historical authenticity is Hecataeus of Miletus, who wrote a book on the popular myths which then controlled the attitude of Greek civilisation. This work, of which only very inconsiderable fragments remain, begins as follows: "Thus speaks Hecataeus of Miletus. I write all this in accordance with what seemed to me to be the truth; for the legends of the Greeks are, in my opinion, contradictory and ridiculous." These words are the cockcrow of individualism, the bugle that sounds the *reveille* of the rationalist faith. Here, for the first time, we have an individual rebelling, in signal isolation, against tradition, that vast millenary world in which the mind of Greece had dwelt from time immemorial.

Reform succeeds reform for a whole century, till we reach the most celebrated innovation of all, that of Cleisthenes. This is how Wilamowitz-Moellendorff outlines the thought and the psychology of his author: "Cleisthenes the Alcmaeonid, belonging to the most powerful of the rival noble families banished by Pisistratus, succeeded, with the help of Delphi and Sparta, in overthrowing the tyrant; he did not, however, take the latter's place, nor did he make Athens an aristocratic state, as Sparta hoped, but, again with the help of Delphi, endowed the city with a fully democratic constitution, the only one we are at all well acquainted with. For it was he, not Solon, who was its true creator . . . Previous sanctions had been confined to unwritten law, religion and custom, but from this date written laws become the true kings. Yet such laws are not dead letters carved in stone, mere obstacles to freedom, but rules of widely accepted validity, such as may be found engraven in the hearts of all enlightened citizens. The people alone have established them; but the people will not cancel them arbitrarily; they must be modified in legal form when they have ceased to be 'just.' The people have appropriated them by the act of pledging their obedience; but it is a legislator who has really made them. In order that the people might be induced to accept them willingly they had to face the same way as the people's thoughts and desires; but it was the legislator who hit upon the creative idea in the course of his self-communings; and just as in the humanitarianism of the old Attic law the mild and pious character of the wise poet, Solon, is clearly to be observed, so in the constitution of Cleisthenes there are traces of a violent type of logico-arithmetical constructive thought which invite the deduction of conclusions as to the temperament of their author. He must have elaborated a complete schematic synthesis of his plans during his banishment, and only admitted with reluctance a few rare compromises with reality when he found he could not extirpate it. His general tendencies, at any rate, have much in common with the arithmetical philosophic speculation which was then beginning and which was soon to lead to the doctrine of faith in the reality of numbers. Cleisthenes had, in fact, connections with Samos, the native city of the Pythagoreans. His violent radicalism derives obviously from the character of the sophists and philosophers, always fanatically determined to impose what is capable of logical proof on the real world in the interests of its salvation. Such castles in the air remind one immediately of the ephemeral constitutions of France prevailing in the interval between the fall of the old monarchy and the rise of Napoleon I."*

I do not think I need add to this exposition. The reform of Cleisthenes is a typically revolutionary phenomenon, the most notable of a long series which comes to an end only with the advent of Pericles. After this date the most casual glance reveals the workings of the geometric mind, philosophic radicalism and "pure reason."

* The resemblance is so close that Cleisthenes, too, introduces the decimal system into his constitution.
U. von Wilamowitz-Moellendorff, *Staat und Gesellschaft der Griechen*.

The purpose of this essay was to show that the genesis of the revolutionary phenomenon must be sought in a determinate affection of the intelligence. Taine brought this idea to light when he enumerated the causes of the great revolution; on the other hand, he cancelled the value of his astute discovery by persuading himself that he was dealing with a habit peculiar to the mentality of France. He did not see that he was dealing with a general historical law. Every people whose development has not been violently interrupted reaches a rationalist stage in the course of its intellectual evolution. When rationalism has been converted into the ordinary method of mental procedure the revolutionary process breaks down automatically and inevitably. It does not, therefore, originate in the oppression of the lower classes by the upper, nor in the advent of an imaginary sensibility to more delicately balanced justice—such a belief is spontaneously rationalist and anti-historical—nor even when new social classes attain sufficient power to wrest supremacy from the hands of its traditional possessors. Certain facts which can be described in this way accompany the manifestation of the revolutionary spirit, but are rather its consequences than its causes.

A beautifully clear proof of this intellectual origin of revolutions may be obtained from the recognition that their radicalism, duration and modulation are proportionate to the nature of the racial intelligence in question. Races which are not particularly intelligent are not particularly revolutionary. The case of Spain is a very clear one: there have been and still are to be found in this country, in great abundance, all the other factors which are usually considered to be decisive in bringing about the explosion of revolution. Nevertheless, the revolutionary spirit proper is still to seek. Our ethnological intelligence has always been an atrophied function and has never had a normal development. The little there has been of subversive temper here could always and can still be reduced to a reflection of that of other countries. Exactly the same is the case with our intelligence: the little there is of that is a reflection of other cultures.

The example of England is very suggestive. It cannot be said that the English people is very intelligent. This is not because they lack intelligence, but because they have no excess of it. They possess the modicum, the amount that is strictly necessary in order to live. For this particular reason their revolutionary era was the most moderate of all and was always tinged with a conservative colouring.

It was the same in Rome. Here, too, was a healthy and virile population, with a great appetite for life and dominion, but not particularly intelligent. Their intellectual awakening comes late and arises in contact with the culture of Greece. The theory I am here advocating has the greatest interest in the questions when the "ideas" of Greece reached Rome and when the revolution commenced. A coincidence of the two dates would have exceptional demonstrative value.

The revolutionary era of Rome begins, as in well-known, in the second century B.C., during the age of the Gracchi.

At that time the typical* situation of Rome is exactly

---

* I should like to be allowed to give the term "typical" its correct meaning. It is generally employed in a sense contrary to that which it ought to bear. By "typical" is usually meant the element differentiating one thing from another, whereas it is rather that which is common to both and corresponds to a "type" or "general class" of things. Thus, physiology is now accustomed to speak of typical digestion—and, in general, of typical functions—which consists in the conjunction of reactions and movements that have to take place in all normal digestion. Every individual, even when normal, adds to that typical process phenomena which are peculiar to himself, but not essential to the digestive function. It is in the same sense that the same as that of Greece between the seventh and sixth centuries and France in the eighteenth. The historical body of Rome has reached the maturity of its inner development; Rome is now what it will be to the end. The first great expansions have begun. Just as Greece annihilated Persia, France and England Spain, so Rome has annihilated the Carthaginian Empire. There is only one difference: the Roman intellect is still rudimentary, rustic, barbarous and medieval. A keen sense for the energetic conduct of practical affairs, coupled with a lack of mental agility, prevent the Roman from feeling that specific enjoyment in the manipulation of ideas which characterizes more intelligent peoples like the Greeks and the French. Up to the epoch of which I am now speaking every purely intellectual occupation had been subjected to furious persecution in Rome. The conventional gesture of hate and scorn of art and thought is destined to endure till the time of Augustus. Even Cicero thinks it necessary to apologise for staying in his villa writing a book instead of attending the Senate.

Such resistance is, however, exercised in vain. The dull and slow intelligence of the agricultural Roman obeys the inexorable cycle and, in its receptive form at any rate, is at last awakened. This phenomenon occurs about 150 B.C. There is then at Rome, for the first time, a select circle of enthusiastic devotees of Greek culture, disdainful of the hostility of the traditionalist masses. The circle is the most illustrious and the highest in social rank in the republic. Scipio Aemilianus, the destroyer of Carthage and Numantia, is the first Roman noble to speak Greek. The historian Polybius and the philosopher Panaetius are his habitual counsellors. At his banquets

---

I refer to the structure of Rome in its essence, whatever its singularities may have been.

the subjects of discussion are poetry, philosophy and the new military technique—for instance, the admirable engineering works which have been revealed by excavations of the Numantine encampments. Just as in Greece the disappearance of the Middle Age coincides with the replacement of the *promachia*, or battle in the form of a series of single combats, by the tactical unit of the phalanx, so in Rome there now begins the organisation of the revolutionary army into cohorts. Marius, the Lafayette of Rome, is to be the actual creator of this innovation.

Scipio is a sentimental adherent of the utopian ideas that have reached him from Greece. There is a tradition that the phrase, *Humanus sum*, which afterwards became *I am man and nothing human is alien from me*, was first heard in his house. Now, that phrase is the eternal motto of cosmopolitan humanitarianism, first invented by Greece and later to be re-invented by the English ideologists and by Voltaire, Diderot and Rousseau. That phrase is, too, the motto of every true revolutionary spirit.

Well, it is in that first "hellenist" and "idealist" circle that the Gracchi, the promoters of the first great revolution, are educated. Their mother, Cornelia, is the mother-in-law and cousin of Scipio Aemilianus.* Tiberius Gracchus had two philosophers as masters and as friends; one was the Greek Diophantes, the other the Italian Blossius: both were fanatical practitioners of political ideology, constructors of utopias. Tiberius, after his fall, proceeded to Asia Minor, where he persuaded the prince, Aristonicus, to employ his slaves and alien settlers in trying the experiment of an utopian state, the City of the Sun,† which resembled the com-

---

* It is well known that Scipio, who belonged to the *gens* Paula Aemiliana, entered the family of the Scipios by adoption.

† Rosenberg: *History of the Roman Republic*, p. 59 (1921).

munal establishment of "phalansteries" advocated by Fourier or the Icaria* of Cabet.

An identical social mechanism, then, is reproduced, and functions through identical devices, in Rome as in Athens and France. The philosopher, the intellectual, is always to be found in the centre of the revolutionary stage. And this is very much to his credit. He is the professional exponent of pure reason, and does his duty in the anti-traditional breach. It may be said that at times when the philosophy of radicalism is in the ascendant—and these are, after all, the most glorious of any historical cycle—the intellectual acquires his maximum power of intervention and his maximum authority. His definitions, his "geometric" concepts, are the explosive substances that time after time in history shatter the cyclopean edifices organised by tradition. So, in our modern Europe, the great French rebellion originated in the abstract definition of man propounded by the encyclopedists. And the latest effort at reconstruction, the doctrine of socialism, arises similarly from the no less abstract definition framed by Marx of man considered simply as a worker, the "pure worker."

In the course of the sunset of revolution ideas gradually cease to be a primary factor in history and return to the negative status they had occupied in the preceding traditionalist age.

* The name given by Cabet to his Utopia. (Translator's note.)

EPILOGUE ON THE MENTAL ATTITUDE OF DISILLUSION

THE theme of the foregoing essay was confined to an attempt at a definition of the revolutionary spirit and an affirmation of its dissolution in Europe. But I said at the beginning of my discourse that such a spirit is a mere stage in the orbit that traverses every great historical cycle. It is preceded by a rationalist attitude and followed by a mystical, or, more precisely, by a superstitious frame of mind. Perhaps the reader feels some curiosity as to the nature of the delta of superstition into which the river of revolution is finally dissipated. It happens, however, that it is not possible to speak upon the subject except at length. The post-revolutionary epochs, after a very fugitive hour of apparent splendour, settle into a time of decadence. And decadences, like births, are enveloped, so far as history is concerned, in darkness and silence. History is accustomed to exercise a strange modesty, which makes it draw a pious veil over the imperfection of commencements and the disagreeable aspect of national decay. It is a fact that the events of the "hellenistic" epoch in Greece and of the middle and later Empire of Rome are little known to historians, while their very existence is scarcely suspected by the generality of educated men. It is not therefore in any way possible to refer to them in the form of a brief allusion.

It would only be by risking the imputation of innumerable misinterpretations that I would venture to satisfy the curiosity of the reader—but are there any curious readers in this country?—in the following words:

The tradionalist mind is a mechanism operating through credulity, for its whole activity consists in its reliance upon the unquestioned wisdom of the past. The rationalist mind breaks these bonds of credulity and replaces them with a fresh imperative: faith in individual energy, of which reason is the supreme instigator. But rationalism tries to do too much—in fact, aspires to the impossible. The proposal to substitute ideas for reality is admirable in its illusive electrical quality, but is always foredoomed to failure. An enterprise so disproportionately ambitious leaves a historical field behind it which becomes an area of disillusion. After the defeat of all his daring idealist aims man is left completely demoralised. He loses all spontaneous faith and does not believe in anything that works along manifest and disciplined lines. He respects neither tradition nor reason, neither collectivity nor the individual. His vital resources weaken because, definitively, it is the beliefs we cherish that keep such resources at concert pitch. He has not sufficient strength in reserve to maintain a suitable attitude before the mystery of life and the universe. Physically and mentally he degenerates. In these epochs the human harvest is left to wither and the national populations dwindle. Not so much through famine, disease or other similar calamities as because the generative potency of man diminishes. Simultaneously, there is a decline in typically virile courage. Universal cowardice begins to prevail: a strange phenomenon which appeared equally in Greece and Rome and has not yet received its due emphasis. In times of security man possesses but half the measure of personal valour required to encounter the vicissitudes of life without disgrace. In such ages of waste valour becomes an unusual quality which is only possessed by a few. Its practice is made a profession whose exponents form a soldiery hostile to all public order and stupidly oppressive of the rest of the social body.

This universal cowardice becomes apparent in the most delicate and intimate recesses of the mind, and projects itself in all directions. Men are terrified once more by lightning and thunder, as they were in the most primitive times. No one relies on his own personal vigour to enable him to triumph over difficulties. Life is felt to be a formidable accident, in which man is dependent upon mysterious and occult wills, acting in accordance with the most puerile caprices. The debased mind is incapable of offering resistance to destiny, and turns to superstitious practices in the hope of propitiating these hidden powers. The most absurd rites attract the adhesion of the multitude. Rome submits to the dominion of all the monstrous divinities of Asia, which had been so honourably disdained two centuries before.

In short: the spirit of the time, being incapable of maintaining itself in equilibrium by its own unaided efforts, searches for some spar that will save it from the wreck, and examines its environment with the anxious and cringing look of a dog, hoping it may find someone to help it. The superstitious mind is, in effect, a dog in search of a master. Men cannot now even remember the noble gestures of pride they once assumed; and the imperative of liberty that resounded in their ears for centuries would now be totally incomprehensible. On the contrary, they feel an incredible anxiety to be slaves. Slavery is their highest ambition: slavery to other men, to an emperor, to a sorcerer or to an idol. Anything rather than feel the terror of facing singlehanded, in their own persons, the ferocious assaults of existence.

Perhaps the name that best suits the spirit that comes into being beyond the sunset of revolution is the term, **spirit of slavery**.

THE HISTORICAL SIGNIFICANCE OF THE THEORY OF EINSTEIN

THE theory of relativity, the most important intellectual fact that the present time can show, inasmuch as it is a theory, admits of discussion whether it is true or false. But, apart from its truth or falsity, a theory is a collection of thoughts which is born in a mind, in a spirit or in a conscience in the same way as a fruit is born upon a tree. Now, a new fruit indicates that a new vegetable species is making its appearance in the flora of the world. Accordingly, we can study the theory of relativity with the same design as a botanist has in describing a plant: we can put aside the question whether the fruit is beneficial or harmful, whether the theory is true or erroneous, and attend solely to the problem of classifying the new species, the new type of living being which we light upon there. Such an analysis will enable us to discover the historical significance of the theory, viz., its nature as an historical phenomenon.

The peculiarities of the theory of relativity point to certain specific tendencies in the mind which has created it. And as a scientific edifice of this magnitude is not the work of one man but the result of the inadvertent collaboration of many, of all the best contemporary minds, in fact, the orientation which these tendencies reveal will indicate the course of western history.

I do not merely mean by this that the triumph of the theory will influence the spirit of mankind by imposing on it the adoption of a definite route. That is an obvious banality. What is really interesting is the inverse proposition: the spirit of man has set out, of its own accord, upon a definite route, and it has therefore been possible for the theory of relativity to be born and to triumph. The more subtle and technical ideas are, the more remote they seem from the ordinary preoccupations of men, the more authentically they denote the profound variations produced in the historical mind of humanity.

It will be enough to lay some little emphasis upon the general tendencies operative in the invention of this theory, and to prolong their lines somewhat beyond the precincts of physics, for the pattern of a new sensibility to shape itself before our eyes, a sensibility antagonistic to that which has prevailed in recent centuries.

1. Absolutism.

The whole system centres, organically, in the idea of relativity. Everything depends, therefore, on the physiognomy assumed by this conception in Einstein's work of genius. It would not be lacking in all sense of proportion to assert that it is at this point that genius has applied its inspired vigour, its thrust of adventurous energy, its sublime archangelic audacity. Once this point was admitted, the rest of the theory could have been worked out with no more than ordinary care.

Classical mechanics recognises the common relativity of all our conclusions on the question of movement and, therefore, the relativity of every position in space and time which the human mind can observe. How is it, then, that the theory of Einstein which, we are told, has destroyed the entire edifice of classical mechanics, throws into relief in its very name, as its principal characteristic, relativity itself? This is the multiform equivocation which we are bound, above all, to expose. *The relativism of Einstein is strictly inverse to that of Galileo and Newton.* For the latter the empirical conclusions we come to concerning duration, location and movement are relative because they believe in the existence of absolute space, time and movement. We cannot perceive them immediately; at most we possess indirect indications of them (centrifugal forces are an example). But if their existence is believed in all the effective conclusions we come to will be disqualified as mere appearances, values relative to the standpoint of comparison occupied by the observer. Consequently, relativism here connotes failure. The physical science of Galileo and Newton is relative in this sense.

Let us suppose that, for one reason or another, a man considers it incumbent upon him to deny the existence of those unattainable absolutes in space, time and transference. At once those concrete conclusions, which formerly appeared relative in the sinister sense of the word, being freed from comparison with the absolute, become the only conclusions that express reality. Absolute (unattainable) reality and a further reality, which is relative in comparison with the former, will not now exist. There will only be one single reality, and this will be what positive physics approximately describes. Now, this reality is what the observer perceives from the place he occupies; it is therefore a relative reality. But as this relative reality, in the suppositious case we have taken, is the only one there is, it must, as well as being relative, be true or, what comes to the same thing, absolute reality. Relativism is not here opposed to absolutism; on the contrary, it merges with it and, so far from suggesting a failure in our knowledge, endows the latter with an absolute validity.

This is the case with the mechanics of Einstein. His physical science is not relative, but relativist, and achieves, thanks to its relativism, an absolute significance.

The most absurd misrepresentation which can be

applied to the new mechanics is to interpret it as one more offspring of the old philosophic relativism, of which it is in fact the executioner. In the old relativism our knowledge is relative because what we aspire to know, viz., space-time reality, is absolute and we cannot attain to it. In the physics of Einstein our knowledge is absolute; it is reality that is relative.

Consequently, we are above all bound to note as one of the most genuine features of the new theory its absolutist tendency in the sphere of knowledge. It is inexplicable that this point should not have been emphasized as a matter of course by those who interpret the philosophic significance of this innovation of genius. The tendency is perfectly clear, however, in the capital formula of the whole theory: physical laws are true whatever may be the system of reference used, that is to say, whatever the point of observation may be. Fifty years ago thinkers were preoccupied with the question whether "from the point of view of Sirius" human truths would be valid. This is equivalent to a degradation of the science practised by man by an attribution to it of a purely domestic value. The mechanics of Einstein permit our physical laws to harmonise with those which may be conjectured to prevail in minds inhabiting Sirius.

But this new absolutism differs radically from that which animated rationalist doctrine during the last few centuries. Such rationalists believed that it was man's privilege to unveil the secrets of nature without doing more than exploring the recesses of his own soul for the eternal truths which it contained. In this belief Descartes creates physics, not from experience, but from what he calls the *trésor de mon esprit*. The value of such truths, which do not proceed from observation, but from pure reason, is universal in character, and instead of learning them from nature we actually, in a certain sense, impose them on nature. They are *a priori* truths. In the works of Newton himself are to be found phrases which reveal the rationalist spirit. "In natural philosophy," he says, "we must abstract our senses." In other words, in order to verify the nature of anything we must turn our backs on it. An example of these magical truths is the law of inertia: according to this law, a moving body, free from all influence, will go on moving indefinitely in a rectilineal and uniform way. Now, such a body, exempt from all influence, is unknown to us. Why make such an affirmation? Simply because space has a rectilineal or euclidian structure and consequently all "spontaneous" movement, which is not diverted by some force, will accommodate itself to the law of space.

But what guarantees this euclidian nature of space? Experience? In a way it does; the nature of pure reason is to resolve, previously to all experience, on the absolute necessity of the space in which physical bodies move being euclidian. Man cannot *see* except in euclidian space. This peculiarity of the inhabitants of the earth is promoted by rationalism to the dignity of a law of the whole cosmos. The old absolutists perpetrated a similar naiveté in every sphere of thought. They begin with an excessive estimate of man. They make him a centre of the universe, though he is only a corner of it. This is the cardinal error that the theory of Einstein now corrects.

2. *Perspectivism.*

The provincial spirit has always, and with good reason, been accused of stupidity. Its nature involves an optical illusion. The provincial does not realise that he is looking at the world from a decentralised position. He supposes, on the contrary, that he is at the centre of the whole earth, and accordingly passes judgment on all things as if his vision were directed from that centre. This is the cause of the deplorable complacency which produces such comic effects. All his opinions are falsified as soon as they are formulated because they originate from a pseudo-centre. On the other hand, the dweller in the capital knows that his city, however large it may be, is only one point of the cosmos, a decentralised corner of it. He knows, further, that the world has no centre, and that it is therefore necessary, in all our judgments, to discount the peculiar perspective that reality offers when it is looked at from our own point of view. This is the reason why the provincial always thinks his neighbour of the great city a sceptic, though the fact is that the latter is only better informed.

The theory of Einstein has shown modern science, with its exemplary discipline—the *nuova scienza* of Galileo, the proud physical philosophy of the West—to have been labouring under an acute form of provincialism. Euclidian geometry, which is only applicable to what is close at hand, had been extended to the whole universe. In Germany to-day the system of Euclid is beginning to be called "proximate geometry" in contradistinction to other collections of axioms which, like those of Riemann, are long-range geometries.

The refutation of this provincial geometry, like that of all provincialism, has been accomplished by means of an apparent limitation, an exercise of modesty in the claims of its conqueror. Einstein is convinced that to talk of Space is a kind of megalomania which inevitably introduces error. We are not aware of any more extensions than those we measure, and we cannot measure more than our instruments can deal with. These are our organ of scientific vision; they determine the spatial structure of the world we know. But as every other being desirous of constructing a system of physics from some other place in the earth is in the same case the result is that there is no real limitation involved at all.

There is no question, then, of our relapsing into a subjectivist interpretation of knowledge, according to which the truth is only true for a pre-determined subjective personality. According to the theory of relativity, the event A, which from the mundane point of view precedes the event B in time, will, from another place in the universe—Sirius, for example—seem to succeed B. There cannot be a more complete inversion of reality. Does it mean that either our own imagination or else that of the mind resident in Sirius is at fault? Not at all. Neither the human mind nor that in Sirius alters the conformation of reality. The fact of the matter is that one of the qualities proper to reality is that of possessing perspective, that is, of organising itself in different ways so as to be visible from different points. Space and time are the objective ingredients of physical perspective, and it is natural that they should vary according to the point of view.

In the introduction to my first *Espectador*, which appeared in January, 1916, when nothing had yet been published on the general theory of relativity,* I put forward a brief exposition of the doctrine of perspective, giving it a range of reference ample enough to transcend physics and include all reality. I mention this fact to show the extent to which a similar cast of thought is a sign of the times.

What surprises me more is that no one has yet noticed this cardinal feature in the work of Einstein. Without a single exception—so far as I know—all that has been written on the matter interprets his great discovery as one step more on the road of subjectivism. In all

* Einstein's first publication on his recent discovery, *Die Grundlagen der allgemeinen Relativitatstheorie* was published in that year.

languages and in all centres of culture we continue to hear that Einstein has confirmed the Kantian doctrine in at least one point, viz., the subjectivity of space and time. It is important for my purpose to declare circumstantially that this belief seems to me the most complete misconception of the significance that the theory of relativity implies.

Let us define the question in a few words, but in the clearest way we can. Perspective is the order and form that reality takes for him who contemplates it. If the place that he occupies varies, the perspective also varies. On the other hand, if another observer is substituted for him in the same place, the perspective remains identical. It is true that if there is no contemplating personality by whom reality is observed there is no perspective. Does this mean that the latter is subjective? Here we have the equivocation which has for centuries, to say the least, misled all philosophy and consequently the attitude of man to the universe. To avoid this difficulty all we have to do is to make a simple distinction.

When we see a stationary and solitary billiard ball we only perceive its qualities of colour and form. But suppose another ball collides with the first. The latter is then driven forward with a speed proportionate to the shock of the collision. Thereupon we note a new quality of the ball, which was previously latent, viz., its resilience. But, someone may say, resilience is not a quality of the first ball, for the quality in question only appears when the second ball collides with it. We shall answer at once that it is not so. Resilience is a quality of the first ball no less than its colour and form, but it is a reactive quality, i.e., one responsive to the action of another object. Thus, in a man, what we usually call his character is his way of reacting to externality—things, persons or events.

Well, now: when some reality collides with another object which we denominate "conscious subject," the reality responds to the subject by *appearing to it*. Appearance is an objective quality of the real, its response to a subject. This response is, moreover, different according to the condition of the observer; for example, according to his standpoint of contemplation. It is to be noted that perspective and point of view now acquire an objective value, though they were previously considered to be deformations imposed by the subject upon reality. Time and space are once more, in defiance of the Kantian thesis, forms of the real.

If there had been among the infinite number of points of view an exceptional one to which it might have been possible to assign a superior correspondence with nature, we could have considered the rest as deforming agents or as "purely subjective." Galileo and Newton believed that this was the case when they spoke of absolute space, that is to say, of a space contemplated from a point of view which is in no way concrete. Newton calls absolute space *sensorium Dei*, the visual organ of God; or, we might say, divine perspective. But we have scarcely thought out in all its implications this idea of a perspective which is not seen from any determined and exclusive place when we discover its contradictory and absurd nature. There is no absolute space because there is no absolute perspective. To be absolute, space has to cease being real—a space full of phenomena—and become an abstraction.

The theory of Einstein is a marvellous proof of the harmonious multiplicity of all possible points of view. If the idea is extended to morals and aesthetics, we shall come to experience history and life in a new way.

The individual who desires to master the maximum amount possible of truth will not now be compelled, as he was for centuries enjoined, to replace his spontaneous point of view with another of an exemplary and standardised character, which used to be called the "vision of things *sub specie aeternitatis*." The point of view of eternity is blind: it sees nothing and does not exist. Man will henceforth endeavour, instead, to be loyal to the unipersonal imperative which represents his individuality.

It is the same with nations. Instead of regarding non-European cultures as barbarous, we shall now begin to respect them, as methods of confronting the cosmos which are equivalent to our own. There is a Chinese perspective which is fully as justified as the Western.

3. Antiutopianism or antirationalism.

The same tendency which in its positive form leads to perspectivism signifies in its negative form hostility to utopianism.

The utopian conception is one which, while believing itself to arise from "nowhere," yet claims to be valid for everyone. To a sensibility of the type evident in the theory of relativity this obstinate refusal to be localised necessarily appears over-confident. There is no spectator of the cosmic spectacle who does not occupy a definite position. To want to see something and not to want to see it from some particular place is an absurdity. Such puerile insubordination to the conditions imposed on us by reality, such incapacity for the cheerful acceptance of destiny, so ingenuous an assumption that it is easy to substitute our own sterile desires, are features of a spirit which is to-day nearing its end and on the verge of giving place to another completely antagonistic to it.

The utopian creed has dominated the European mind during the whole of the modern epoch in science, in morals, in religion and in art. The whole weight of the intensely earnest desire to master reality—a specifically European characteristic—had to be thrown into the scales to prevent Western civilisation from perishing in a gigantic fiasco. For the most troublesome feature of utopianism is not that it gives us false solutions to problems—scientific or political—but something worse: the difficulty is that it does not accept the problem of the real as it is presented, but immediately, viz., *a priori*, imposes a form on it which is capricious.

If we compare Western life with that of Asia—Indian or Chinese—we are at once struck by the spiritual instability of the European as opposed to the profound equilibrium of the Oriental mind. This equilibrium reveals the fact that, at any rate in the greatest problems of life, the Easterner has discovered formulae more perfectly adjusted to reality. The European, on the other hand, has been frivolous in his appreciation of the elemental factors of life and has contrived capricious interpretations of them which have periodically to be replaced.

The utopist aberration of human intelligence begins in Greece and occurs wherever rationalism reaches the point of exacerbation. Pure reason constructs an exemplary world—a physical or political cosmos—in the belief that it is the true reality and must therefore supplant the actually existent one. The divergence between phenomena and pure ideas is such that the conflict is inevitable. But the rationalist is sure that the struggle will result in the defeat of reality. *This conviction is the main characteristic of the rationalist temperament.*

Reality, naturally, possesses more than sufficient toughness to resist the assaults of ideas. Rationalism then looks for a way out: it recognises that, *for the moment*, the idea cannot be realised, but believes that success will be achieved in an "infinite process" (Leibnitz, Kant).

Utopianism takes the form of "uchronianism."* During the last two centuries and a half every difficulty was resolved by an appeal to the infinite, or at least to periods of indeterminate length. In Darwinism, for instance, one species is born of another without the intervention of more than a few millennia between the two. It is assumed that time, that ghostly river, by merely elapsing, can be an efficient cause and make what is actually inconceivable a probability.

We do not realise that science, whose sole pleasure is to obtain a reliable image of nature, can be nourished on illusion. I remember one detail that has exercised a very great deal of influence over my thought. Many years ago I was reading a lecture of the physiologist Loeb on tropism, a concept by means of which it was thought possible to describe and explain the law which regulates the elemental movements of infusoria. This concept, with certain corrections and additions, serves as a basis for understanding some of these phenomena. But at the end of his lecture Loeb adds: "The time will come when what we call to-day the moral acts of man will be explained simply as tropisms." This piece of audacity shocked me extremely, for it opened my eyes to many other opinions of modern science which make, with less ostentation, the same mistake. So then, I thought, such a concept as tropism, which is scarcely capable of penetrating the secret of phenomena so simple as the transference of infusoria, can be thought sufficient, in some vague future, to explain so mysterious and complex a thing as the ethical acts of man. What sense can there be in this? Science has to solve its problems to-day, not put us off to the Greek kalends. If its actual methods are not enough at present to master the riddle of the

---

Ucronismo, a word coined from the Greek, on the analogy of *utopismo*, substituting "time" for "place." (Translator's note).

universe the proper thing to do is to replace them with others which may be more efficacious. Current science, however, is full of problems which are left intact because they are incompatible with the methods employed. It is, apparently, the former which are to overcome the latter, and not *vice versa*. Science is full of uchronianism, of Greek kalends.

When we emerge from this scientific beatitude, with its cult of idolatrous worship of pre-established methods, and turn to the thought of Einstein we feel, as it were, a fresh morning breeze. The attitude of Einstein is completely distinct from the traditional one. We see him advancing directly upon problems with the gestures of a young athlete and, by employing the method readiest to hand, catching them by the horns. He makes a virtue and an efficacious system of tactics out of what appeared to be a defect and a limitation in science.

A short digression will enable us to see this question in a clearer light.

One part of the work of Kant will remain imperishable, viz., his great discovery that experience is not only the aggregate of data transmitted by the senses, but also a product of two factors. The sensible datum has to be received, given its correct affiliation and organised in a system of disposition. This order is supplied by the subjective personality and is *a priori*. In other words, physical experience is a compound of observation and geometry. Geometry is a pentagraph elaborated by pure reason: observation is the work of the senses. All science which is explanatory of material phenomena has contained, contains and will contain these two ingredients.

This identity of composition, invariably exhibited by modern physics throughout its entire history, does not, however, exclude the most profound variations in its spirit. The mutual relation maintained between its two ingredients leaves room, in fact, for very diverse interpretations. Which of the two is to supplant the other? Ought observation to yield to the demands of geometry, or geometry to observation? To decide one way or the other will mean our adherence to one of two antagonistic types of intellectual tendency. There is room for two opposed castes of opinion in one and the same system of physics.

It is common knowledge that the experiment of Michelson* is crucial in the hierarchy of such tests: physical theory is there placed between the devil and the deep sea. The geometrical law which proclaims the unalterable homogeneity of space, whatever may be the processes which occur in it, enters into uncompromising conflict with observation, with fact, with matter. One of two things must happen: either matter is to yield to geometry or the latter to the former.

In this acute dilemma two intellectual temperaments come before us, and we are able to observe their reaction. Lorentz and Einstein, confronted by the same experiment, take opposite resolutions. Lorentz, in this particular representing the old rationalism, believes himself obliged to conclude that it is matter which yields and contracts. The celebrated "contradiction of Lorentz" is an admirable example of utopianism. It is the Oath of the Tennis Court transferred to physics. Einstein adopts the contrary solution. Geometry must yield, pure space is to bow to observation, to curve, in fact.

In the political sphere, supposing the analogy to be a perfect one, Lorentz would say: Nations may perish, provided we keep our principles. Einstein, on the other

---

* The German physicist, born in 1852. His experiment dealt with the propagation of light as observed on the earth. (Translator's note.)

hand, would maintain: We must look for such principles as will preserve nations, because that is what principles are for.

It is not easy to exaggerate the importance of the change of course imposed by Einstein upon physical science. Hitherto the *rôle* of geometry, of pure reason, has been to exercise an undisputed dictatorship. Common speech retains a trace of the sublime function which used to be attributed to reason: people talk of the "dictates of reason." For Einstein the *rôle* of reason is a much more modest one: it descends from dictatorship to the status of a humble instrument, which has, in every case, first to prove its efficiency.

Galileo and Newton made the universe euclidian simply because reason dictated it so. But pure reason cannot do anything but invent systems of methodical arrangement. These may be very numerous and various. Euclidian geometry is one, Riemann's another, Lobatchewski's another, and so on. But it is clearly not these systems, not pure reason, which resolve the nature of the real. On the contrary, reality selects from among these possible orders or schemes the one which has most affinity with itself. This is what the theory of relativity means. The rationalist past of four centuries is confronted by the genius of Einstein, who inverts the time-honoured relation which used to exist between reason and observation. Reason ceases to be an imperative standard and is converted into an arsenal of instruments; observation tests these and decides which is the most convenient to use. The result is the creation of the science of mutual selection between pure ideas and pure facts.

This is one of the features which it is most important to emphasize in the thought of Einstein, for here we discover the initiation of an entirely new attitude to life.

Culture ceases to be, as hitherto, an imperative standard to which our existence has to conform. We can now see a more delicate and more just relation between the two factors. Certain phenomena of life are selected as possible forms of culture; but of these possible forms of culture life, in its turn, selects the only ones which are suitable for future realisation.

4. *Finitism.*

I should not like to conclude this genealogical sketch of the profound tendencies rife in the theory of relativity without alluding to the most clear and patent of them. While the utopist past used to settle all disputes by the expedient of recourse to the infinite in space and time, the physics of Einstein—and similarly the recent mathematical systems of Brouwer and Wey—annotates the universe. The world of Einstein is curved, and therefore closed and finite.*

For anyone who believes that scientific doctrines are born by means of spontaneous generation and need do no more than open our eyes and minds to facts the innovation under discussion has no real importance. It merely amounts to a modification of the form which used to be attributed to the world. But the original supposition is false: a scientific doctrine is not born, however obvious the facts upon which it is based may appear, without a well-defined spiritual orientation. It is necessary to understand the genesis of our thoughts in all their delicate duplicity. No more truths are discovered than those we are already in search of. To the rest, however evident they may be, the spirit is blind.

* The system of Einstein prosecutes its attack on the infinite in all directions. For example, it excludes the possibility of infinite velocities.

This gives an enormous range of reference to the fact that physics and mathematics are suddenly beginning to have a marked preference for the finite and a great distaste for the infinite. Can there be a more radical difference between two minds than that one should tend to the idea that the universe is unlimited and that the other should feel its environment to be circumscribed? The infinity of the cosmos was one of the great intoxicating ideas produced by the Renaissance. It flooded the hearts of men with tides of pathetic emotion, and Giordano Bruno suffered a cruel death on its behalf. During the whole of the modern epoch the most earnest desires of Western man have concealed, as though it were a magical foundation for them, this idea of the infinity of the cosmic scene.

And now, all at once, the world has become limited, a garden surrounded by confining walls, an apartment, an interior. Does not this new setting suggest an entirely different style of living, altogether opposed to that at present in use? Our grandsons will enter existence armed with this notion, and their attitude to space will have a meaning contrary to that of our own. There is evident in this propensity to finitism a definite urge towards limitation, towards beauty of serene type, towards antipathy to vague superlatives, towards antiromanticism. The Greek, the "classical" man, also lived in a limited universe. All Greek culture has a horror of the infinite and seeks the *metron*, the mean.

It would be superficial, however, to believe that the human mind is being directed towards a new classicism. There has never yet been a new classicism which has not resulted in frivolity. The classical man seeks the limit, but it is because he has never lived in an unlimited world. Our case is inverse: the limit signifies an amputation for us, and the closed and finite world in which we are now to draw breath will be, irremediably, a truncated universe.*

* It would be necessary to touch on two other points to complete the general outline of the mind which has created the theory of relativity. One of them would be the care with which the discontinuities in the real are emphasised, as opposed to the passion for the continuous which dominates the thought of the last few centuries. This *discontinuism* is equally triumphant in biology and in history. The other point, perhaps the most weighty of all, would be the tendency to suppress causality, which operates in a latent form in the theory of Einstein. Physics, which began by being *mechanics* and then became *dynamics*, tends in Eisntein to be converted into mere *cinematics*. On both points it is only possible to speak by reference to difficult technical questions which I have tried to eliminate in the present text.

Revised October 31, 1965

harper ⚜ torchbooks

HUMANITIES AND SOCIAL SCIENCES

American Studies: General

THOMAS C. COCHRAN: The Inner Revolution: *Essays on the Social Sciences in History* TB/1140
EDWARD S. CORWIN: American Constitutional History. *Essays edited by Alpheus T. Mason and Gerald Garvey* TB/1136
A. HUNTER DUPREE: Science in the Federal Government: *A History of Policies and Activities to 1940* TB/573
OSCAR HANDLIN, Ed.: This Was America: *As Recorded by European Travelers in the Eighteenth, Nineteenth and Twentieth Centuries. Illus.* TB/1119
MARCUS LEE HANSEN: The Atlantic Migration: 1607-1860. *Edited by Arthur M. Schlesinger; Introduction by Oscar Handlin* TB/1052
MARCUS LEE HANSEN: The Immigrant in American History. *Edited with a Foreword by Arthur M. Schlesinger* TB/1120
JOHN HIGHAM, Ed.: The Reconstruction of American History TB/1068
ROBERT H. JACKSON: The Supreme Court in the American System of Government TB/1106
JOHN F. KENNEDY: A Nation of Immigrants. *Illus. Revised and Enlarged. Introduction by Robert F. Kennedy* TB/1118
RALPH BARTON PERRY: Puritanism and Democracy TB/1138
ARNOLD ROSE: The Negro in America: *The Condensed Version of Gunnar Myrdal's An American Dilemma* TB/3048
MAURICE R. STEIN: The Eclipse of Community: *An Interpretation of American Studies* TB/1128
W. LLOYD WARNER and Associates: Democracy in Jonesville: *A Study in Quality and Inequality* ‖ TB/1129
W. LLOYD WARNER: Social Class in America: *The Evaluation of Status* TB/1013

American Studies: Colonial

BERNARD BAILYN, Ed.: The Apologia of Robert Keayne: *Self-Portrait of a Puritan Merchant* TB/1201
BERNARD BAILYN: The New England Merchants in the Seventeenth Century TB/1149
JOSEPH CHARLES: The Origins of the American Party System TB/1049
LAWRENCE HENRY GIPSON: The Coming of the Revolution: 1763-1775. † *Illus.* TB/3007

LEONARD W. LEVY: Freedom of Speech and Press in Early American History: *Legacy of Suppression* TB/1109
PERRY MILLER: Errand Into the Wilderness TB/1139
PERRY MILLER & T. H. JOHNSON, Eds.: The Puritans: *A Sourcebook of Their Writings* Vol. I TB/1093; Vol. II TB/1094
KENNETH B. MURDOCK: Literature and Theology in Colonial New England TB/99
WALLACE NOTESTEIN: The English People on the Eve of Colonization: 1603-1630. † *Illus.* TB/3006
LOUIS B. WRIGHT: The Cultural Life of the American Colonies: 1607-1763. † *Illus.* TB/3005

American Studies: From the Revolution to the Civil War

JOHN R. ALDEN: The American Revolution: 1775-1783. † *Illus.* TB/3011
RAY A. BILLINGTON: The Far Western Frontier: 1830-1860. † *Illus.* TB/3012
GEORGE DANGERFIELD: The Awakening of American Nationalism: 1815-1828. † *Illus.* TB/3061
CLEMENT EATON: The Freedom-of-Thought Struggle in the Old South. *Revised and Enlarged. Illus.* TB/1150
CLEMENT EATON: The Growth of Southern Civilization: 1790-1860. † *Illus.* TB/3040
LOUIS FILLER: The Crusade Against Slavery: 1830-1860. † *Illus.* TB/3029
DIXON RYAN FOX: The Decline of Aristocracy in the Politics of New York: 1801-1840. ‡ *Edited by Robert V. Remini* TB/3064
FELIX GILBERT: The Beginnings of American Foreign Policy: *To the Farewell Address* TB/1200
FRANCIS J. GRUND: Aristocracy in America: *Social Class in the Formative Years of the New Nation* TB/1001
ALEXANDER HAMILTON: The Reports of Alexander Hamilton. ‡ *Edited by Jacob E. Cooke* TB/3060
THOMAS JEFFERSON: Notes on the State of Virginia. ‡ *Edited by Thomas P. Abernethy* TB/3052
BERNARD MAYO: Myths and Men: *Patrick Henry, George Washington, Thomas Jefferson* TB/1108
JOHN C. MILLER: Alexander Hamilton and the Growth of the New Nation TB/1057
RICHARD B. MORRIS, Ed.: The Era of the American Revolution TB/1180
R. B. NYE: The Cultural Life of the New Nation: 1776-1801. † *Illus.* TB/3026

† The New American Nation Series, edited by Henry Steele Commager and Richard B. Morris.
‡ American Perspectives series, edited by Bernard Wishy and William E. Leuchtenburg.
* The Rise of Modern Europe series, edited by William L. Langer.
‖ Researches in the Social, Cultural, and Behavioral Sciences, edited by Benjamin Nelson.
§ The Library of Religion and Culture, edited by Benjamin Nelson.
ᵒ Not for sale in Canada.
Σ Harper Modern Science Series, edited by James R. Newman.

1

FRANK THISTLETHWAITE: America and the Atlantic Community: *Anglo-American Aspects, 1790-1850* TB/1107
A. F. TYLER: Freedom's Ferment: *Phases of American Social History from the Revolution to the Outbreak of the Civil War. 31 illus.* TB/1074
GLYNDON G. VAN DEUSEN: The Jacksonian Era: 1828-1848. † *Illus.* TB/3028
LOUIS B. WRIGHT: Culture on the Moving Frontier TB/1053

American Studies: Since the Civil War

RAY STANNARD BAKER: Following the Color Line: *American Negro Citizenship in Progressive Era.* ‡ *Illus.* Edited by Dewey W. Grantham, Jr. TB/3053
RANDOLPH S. BOURNE: War and the Intellectuals: *Collected Essays, 1915-1919.* ‡ Ed. by Carl Resek TB/3043
A. RUSSELL BUCHANAN: The United States and World War II. † *Illus.* Vol. I TB/3044; Vol. II TB/3045
ABRAHAM CAHAN: The Rise of David Levinsky: *a documentary novel of social mobility in early twentieth century America.* Intro. by John Higham TB/1028
THOMAS C. COCHRAN: The American Business System: *A Historical Perspective, 1900-1955* TB/1080
THOMAS C. COCHRAN & WILLIAM MILLER: The Age of Enterprise: *A Social History of Industrial America* TB/1054
FOSTER RHEA DULLES: America's Rise to World Power: 1898-1954. † *Illus.* TB/3021
W. A. DUNNING: Essays on the Civil War and Reconstruction. *Introduction by David Donald* TB/1181
W. A. DUNNING: Reconstruction, Political and Economic: 1865-1877 TB/1073
HAROLD U. FAULKNER: Politics, Reform and Expansion: 1890-1900. † *Illus.* TB/3020
JOHN D. HICKS: Republican Ascendancy: 1921-1933. † *Illus.* TB/3041
ROBERT HUNTER: Poverty: *Social Conscience in the Progressive Era.* ‡ Edited by Peter d'A. Jones TB/3065
HELEN HUNT JACKSON: A Century of Dishonor: *The Early Crusade for Indian Reform.* ‡ Edited by Andrew F. Rolle TB/3063
ALBERT D. KIRWAN: Revolt of the Rednecks: *Mississippi Politics, 1876-1925* TB/1199
WILLIAM L. LANGER & S. EVERETT GLEASON: The Challenge to Isolation: *The World Crisis of 1937-1940 and American Foreign Policy* Vol. I TB/3054; Vol. II TB/3055
WILLIAM E. LEUCHTENBURG: Franklin D. Roosevelt and the New Deal: 1932-1940. † *Illus.* TB/3025
ARTHUR S. LINK: Woodrow Wilson and the Progressive Era: 1910-1917. † *Illus.* TB/3023
ROBERT GREEN MCCLOSKEY: American Conservatism in the Age of Enterprise: 1865-1910 TB/1137
GEORGE E. MOWRY: The Era of Theodore Roosevelt and the Birth of Modern America: 1900-1912. † *Illus.* TB/3022
RUSSEL B. NYE: Midwestern Progressive Politics: *A Historical Study of its Origins and Development, 1870-1958* TB/1202
WALTER RAUSCHENBUSCH: Christianity and the Social Crisis. ‡ Edited by Robert D. Cross TB/3059
WHITELAW REID: After the War: *A Tour of the Southern States, 1865-1866.* ‡ Edited by C. Vann Woodward TB/3066
CHARLES H. SHINN: Mining Camps: *A Study in American Frontier Government.* ‡ Edited by Rodman W. Paul TB/3062
TWELVE SOUTHERNERS: I'll Take My Stand: *The South and the Agrarian Tradition.* Intro. by Louis D. Rubin, Jr.; Biographical Essays by Virginia Rock TB/1072

WALTER E. WEYL: The New Democracy: *An Essay on Certain Political Tendencies in the United States.* ‡ Edited by Charles B. Forcey TB/3042
VERNON LANE WHARTON: The Negro in Mississippi: 1865-1890 TB/1178

Anthropology

JACQUES BARZUN: Race: *A Study in Superstition.* Revised Edition TB/1172
JOSEPH B. CASAGRANDE, Ed.: In the Company of Man: *Twenty Portraits of Anthropological Informants. Illus.* TB/3047
W. E. LE GROS CLARK: The Antecedents of Man: *Intro. to Evolution of the Primates.* º *Illus.* TB/559
CORA DU BOIS: The People of Alor. *New Preface by the author. Illus.* Vol. I TB/1042; Vol. II TB/1043
RAYMOND FIRTH, Ed.: Man and Culture: *An Evaluation of the Work of Bronislaw Malinowski* ‖ º TB/1133
L. S. B. LEAKEY: Adam's Ancestors: *The Evolution of Man and His Culture. Illus.* TB/1019
ROBERT H. LOWIE: Primitive Society. *Introduction by Fred Eggan* TB/1056
SIR EDWARD TYLOR: The Origins of Culture. *Part I of "Primitive Culture."* § *Intro. by Paul Radin* TB/33
SIR EDWARD TYLOR: Religion in Primitive Culture. *Part II of "Primitive Culture."* § *Intro. by Paul Radin* TB/34
W. LLOYD WARNER: A Black Civilization: *A Study of an Australian Tribe.* ‖ *Illus.* TB/3056

Art and Art History

WALTER LOWRIE: Art in the Early Church. *Revised Edition. 452 illus.* TB/124
EMILE MÂLE: The Gothic Image: *Religious Art in France of the Thirteenth Century.* § *190 illus.* TB/44
MILLARD MEISS: Painting in Florence and Siena after the Black Death: *The Arts, Religion and Society in the Mid-Fourteenth Century. 169 illus.* TB/1148
ERICH NEUMANN: The Archetypal World of Henry Moore. *107 illus.* TB/2020
DORA & ERWIN PANOFSKY: Pandora's Box: *The Changing Aspects of a Mythical Symbol. Revised Edition. Illus.* TB/2021
ERWIN PANOFSKY: Studies in Iconology: *Humanistic Themes in the Art of the Renaissance. 180 illustrations* TB/1077
ALEXANDRE PIANKOFF: The Shrines of Tut-Ankh-Amon. *Edited by N. Rambova. 117 illus.* TB/2011
JEAN SEZNEC: The Survival of the Pagan Gods: *The Mythological Tradition and Its Place in Renaissance Humanism and Art. 108 illustrations* TB/2004
OTTO VON SIMSON: The Gothic Cathedral: *Origins of Gothic Architecture and the Medieval Concept of Order. 58 illus.* TB/2018
HEINRICH ZIMMER: Myth and Symbols in Indian Art and Civilization. *70 illustrations* TB/2005

Business, Economics & Economic History

REINHARD BENDIX: Work and Authority in Industry: *Ideologies of Management in the Course of Industrialization* TB/3035
GILBERT BURCK & EDITORS OF FORTUNE: The Computer Age: *And Its Potential for Management* TB/1179
THOMAS C. COCHRAN: The American Business System: *A Historical Perspective, 1900-1955* TB/1080
THOMAS C. COCHRAN: The Inner Revolution: *Essays on the Social Sciences in History* TB/1140

THOMAS C. COCHRAN & WILLIAM MILLER: The Age of Enterprise: *A Social History of Industrial America* TB/1054

ROBERT DAHL & CHARLES E. LINDBLOM: Politics, Economics, and Welfare: *Planning & Politico-Economic Systems Resolved into Basic Social Processes* TB/3037

PETER F. DRUCKER: The New Society: *The Anatomy of Industrial Order* TB/1082

EDITORS OF FORTUNE: America in the Sixties: *The Economy and the Society* TB/1015

ROBERT L. HEILBRONER: The Great Ascent: *The Struggle for Economic Development in Our Time* TB/3030

FRANK H. KNIGHT: The Economic Organization TB/1214

FRANK H. KNIGHT: Risk, Uncertainty and Profit TB/1215

ABBA P. LERNER: Everybody's Business: *Current Assumptions in Economics and Public Policy* TB/3051

ROBERT GREEN MC CLOSKEY: American Conservatism in the Age of Enterprise, 1865-1910 TB/1137

PAUL MANTOUX: The Industrial Revolution in the Eighteenth Century: *The Beginnings of the Modern Factory System in England* ⁰ TB/1079

WILLIAM MILLER, Ed.: Men in Business: *Essays on the Historical Role of the Entrepreneur* TB/1081

PERRIN STRYKER: The Character of the Executive: *Eleven Studies in Managerial Qualities* TB/1041

PIERRE URI: Partnership for Progress: *A Program for Transatlantic Action* TB/3036

Contemporary Culture

JACQUES BARZUN: The House of Intellect TB/1051

JOHN U. NEF: Cultural Foundations of Industrial Civilization TB/1024

NATHAN M. PUSEY: The Age of the Scholar: *Observations on Education in a Troubled Decade* TB/1157

PAUL VALÉRY: The Outlook for Intelligence TB/2016

Historiography & Philosophy of History

JACOB BURCKHARDT: On History and Historians. *Intro. by H. R. Trevor-Roper* TB/1216

WILHELM DILTHEY: Pattern and Meaning in History: *Thoughts on History and Society.* ⁰ Edited with an Introduction by H. P. Rickman TB/1075

H. STUART HUGHES: History as Art and as Science: *Twin Vistas on the Past* TB/1207

RAYMOND KLIBANSKY & H. J. PATON, Eds.: Philosophy and History: *The Ernst Cassirer Festschrift. Illus.* TB/1115

GEORGE N. NADEL, Ed.: Studies in the Philosophy of History: *Selected Essays from* History and Theory TB/1208

JOSE ORTEGA Y GASSET: The Modern Theme. *Introduction by Jose Ferrater Mora* TB/1038

SIR KARL POPPER: The Open Society and Its Enemies
Vol. I: *The Spell of Plato* TB/1101
Vol. II: *The High Tide of Prophecy: Hegel, Marx and the Aftermath* TB/1102

SIR KARL POPPER: The Poverty of Historicism ⁰ TB/1126

G. J. RENIER: History: Its Purpose and Method TB/1209

W. H. WALSH: Philosophy of History: *An Introduction* TB/1020

History: General

L. CARRINGTON GOODRICH: A Short History of the Chinese People. *Illus.* TB/3015

DAN N. JACOBS & HANS H. BAERWALD: Chinese Communism: *Selected Documents* TB/3031

BERNARD LEWIS: The Arabs in History TB/1029

SIR PERCY SYKES: A History of Exploration. ⁰ *Introduction by John K. Wright* TB/1046

History: Ancient and Medieval

A. ANDREWES: The Greek Tyrants TB/1103

P. BOISSONNADE: Life and Work in Medieval Europe: *The Evolution of the Medieval Economy, the 5th to the 15th Century.* ⁰ Preface by Lynn White, Jr. TB/1141

HELEN CAM: England before Elizabeth TB/1026

NORMAN COHN: The Pursuit of the Millennium: *Revolutionary Messianism in Medieval and Reformation Europe* TB/1037

G. G. COULTON: Medieval Village, Manor, and Monastery TB/1022

HEINRICH FICHTENAU: The Carolingian Empire: *The Age of Charlemagne* TB/1142

F. L. GANSHOF: Feudalism TB/1058

EDWARD GIBBON: The Triumph of Christendom in the Roman Empire *(Chaps. XV-XX of "Decline and Fall," J. B. Bury edition).* § Illus. TB/46

MICHAEL GRANT: Ancient History ⁰ TB/1190

W. O. HASSALL, Ed.: Medieval England: *As Viewed by Contemporaries* TB/1205

DENYS HAY: The Medieval Centuries ⁰ TB/1192

J. M. HUSSEY: The Byzantine World TB/1057

SAMUEL NOAH KRAMER: Sumerian Mythology TB/1055

FERDINAND LOT: The End of the Ancient World and the Beginnings of the Middle Ages. *Introduction by Glanville Downey* TB/1044

G. MOLLATT: The Popes at Avignon: 1305-1378 TB/308

CHARLES PETIT-DUTAILLIS: The Feudal Monarchy in France and England: *From the Tenth to the Thirteenth Century* ⁰ TB/1165

HENRI PIERENNE: Early Democracies in the Low Countries: *Urban Society and Political Conflict in the Middle Ages and the Renaissance. Introduction by John H. Mundy* TB/1110

STEVEN RUNCIMAN: A History of the Crusades. Volume I: *The First Crusade and the Foundation of the Kingdom of Jerusalem. Illus.* TB/1143

FERDINAND SCHEVILL: Siena: *The History of a Medieval Commune. Intro. by William M. Bowsky* TB/1164

SULPICIUS SEVERUS et al.: The Western Fathers: *Being the Lives of Martin of Tours, Ambrose, Augustine of Hippo, Honoratus of Arles and Germanus of Auxerre. Edited and translated by F. R. Hoare* TB/309

HENRY OSBORN TAYLOR: The Classical Heritage of the Middle Ages. *Foreword and Biblio. by Kenneth M. Setton* TB/1117

F. VAN DER MEER: Augustine The Bishop: *Church and Society at the Dawn of the Middle Ages* TB/304

J. M. WALLACE-HADRILL: The Barbarian West: *The Early Middle Ages, A.D. 400-1000* TB/1061

History: Renaissance & Reformation

JACOB BURCKHARDT: The Civilization of the Renaissance in Italy. *Intro. by Benjamin Nelson & Charles Trinkaus. Illus.* Vol. I TB/40; Vol. II TB/41

ERNST CASSIRER: The Individual and the Cosmos in Renaissance Philosophy. *Translated with an Introduction by Mario Domandi* TB/1097

FEDERICO CHABOD: Machiavelli and the Renaissance TB/1193

EDWARD P. CHEYNEY: The Dawn of a New Era, 1250-1453. * *Illus.* TB/3002

R. TREVOR DAVIES: The Golden Century of Spain, 1501-1621 ⁰ TB/1194

DESIDERIUS ERASMUS: Christian Humanism and the Reformation: *Selected Writings. Edited and translated by John C. Olin* TB/1166

3

WALLACE K. FERGUSON et al.: Facets of the Renaissance TB/1098
WALLACE K. FERGUSON et al.: The Renaissance: Six Essays. Illus. TB/1084
JOHN NEVILLE FIGGIS: The Divine Right of Kings. Introduction by G. R. Elton TB/1191
JOHN NEVILLE FIGGIS: Political Thought from Gerson to Grotius: 1414-1625: Seven Studies. Introduction by Garrett Mattingly TB/1032
MYRON P. GILMORE: The World of Humanism, 1453-1517.* Illus. TB/3003
FRANCESCO GUICCIARDINI: Maxims and Reflections of a Renaissance Statesman (Ricordi). Trans. by Mario Domandi. Intro. by Nicolai Rubinstein TB/1160
J. H. HEXTER: More's Utopia: The Biography of an Idea New Epilogue by the Author TB/1195
JOHAN HUIZINGA: Erasmus and the Age of Reformation. Illus. TB/19
ULRICH VON HUTTEN et al.: On the Eve of the Reformation: "Letters of Obscure Men." Introduction by Hajo Holborn TB/1124
PAUL O. KRISTELLER: Renaissance Thought: The Classic, Scholastic, and Humanist Strains TB/1048
PAUL O. KRISTELLER: Renaissance Thought II: Papers on Humanism and the Arts TB/1163
NICCOLO MACHIAVELLI: History of Florence and of the Affairs of Italy: from the earliest times to the death of Lorenzo the Magnificent. Introduction by Felix Gilbert TB/1027
ALFRED VON MARTIN: Sociology of the Renaissance. Introduction by Wallace K. Ferguson TB/1099
GARRETT MATTINGLY et al.: Renaissance Profiles. Edited by J. H. Plumb TB/1162
MILLARD MEISS: Painting in Florence and Siena after the Black Death: The Arts, Religion and Society in the Mid-Fourteenth Century. 169 illus. TB/1148
J. E. NEALE: The Age of Catherine de Medici º TB/1085
ERWIN PANOFSKY: Studies in Iconology: Humanistic Themes in the Art of the Renaissance. 180 illustrations TB/1077
J. H. PARRY: The Establishment of the European Hegemony: 1415-1715: Trade and Exploration in the Age of the Renaissance TB/1045
J. H. PLUMB: The Italian Renaissance: A Concise Survey of Its History and Culture TB/1161
CECIL ROTH: The Jews in the Renaissance. Illus. TB/834
GORDON RUPP: Luther's Progress to the Diet of Worms º TB/120
FERDINAND SCHEVILL: The Medici. Illus. TB/1010
FERDINAND SCHEVILL: Medieval and Renaissance Florence. Illus. Volume I: Medieval Florence TB/1090 Volume II: The Coming of Humanism and the Age of the Medici TB/1091
G. M. TREVELYAN: England in the Age of Wycliffe, 1368-1520 º TB/1112
VESPASIANO: Renaissance Princes, Popes, and Prelates: The Vespasiano Memoirs: Lives of Illustrious Men of the XVth Century. Intro. by Myron P. Gilmore TB/1111

History: Modern European

FREDERICK B. ARTZ: Reaction and Revolution, 1815-1832. * Illus. TB/3034
MAX BELOFF: The Age of Absolutism, 1660-1815 TB/1062
ROBERT C. BINKLEY: Realism and Nationalism, 1852-1871. * Illus. TB/3038
ASA BRIGGS: The Making of Modern England, 1784-1867: The Age of Improvement º TB/1203
CRANE BRINTON: A Decade of Revolution, 1789-1799. * Illus. TB/3018

J. BRONOWSKI & BRUCE MAZLISH: The Western Intellectual Tradition: From Leonardo to Hegel TB/3001
GEOFFREY BRUUN: Europe and the French Imperium, 1799-1814. * Illus. TB/3033
ALAN BULLOCK: Hitler, A Study in Tyranny. º Illus. TB/1123
E. H. CARR: The Twenty Years' Crisis, 1919-1939: An Introduction to the Study of International Relations º TB/1122
GORDON A. CRAIG: From Bismarck to Adenauer: Aspects of German Statecraft. Revised Edition TB/1171
WALTER L. DORN: Competition for Empire, 1740-1763. * Illus. TB/3032
CARL J. FRIEDRICH: The Age of the Baroque, 1610-1660. * Illus. TB/3004
RENÉ FUELOEP-MILLER: The Mind and Face of Bolshevism: An Examination of Cultural Life in Soviet Russia. New Epilogue by the Author TB/1188
M. DOROTHY GEORGE: London Life in the Eighteenth Century TB/1182
LEO GERSHOY: From Despotism to Revolution, 1763-1789. * Illus. TB/3017
C. C. GILLISPIE: Genesis and Geology: The Decades before Darwin § TB/51
ALBERT GOODWIN: The French Revolution TB/1064
ALBERT GUERARD: France in the Classical Age: The Life and Death of an Ideal TB/1183
CARLTON J. H. HAYES: A Generation of Materialism, 1871-1900. * Illus. TB/3039
J. H. HEXTER: Reappraisals in History: New Views on History & Society in Early Modern Europe TB/1100
A. R. HUMPHREYS: The Augustan World: Society, Thought, and Letters in 18th Century England º TB/1105
ALDOUS HUXLEY: The Devils of Loudun: A Study in the Psychology of Power Politics and Mystical Religion in the France of Cardinal Richelieu § º TB/60
DAN N. JACOBS, Ed.: The New Communist Manifesto & Related Documents. Third edition, revised TB/1078
HANS KOHN: The Mind of Germany: The Education of a Nation TB/1204
HANS KOHN, Ed.: The Mind of Modern Russia: Historical and Political Thought of Russia's Great Age TB/1065
KINGSLEY MARTIN: French Liberal Thought in the Eighteenth Century: A Study of Political Ideas from Bayle to Condorcet TB/1114
SIR LEWIS NAMIER: Personalities and Powers: Selected Essays TB/1186
SIR LEWIS NAMIER: Vanished Supremacies: Essays on European History, 1812-1918 º TB/1088
JOHN U. NEF: Western Civilization Since the Renaissance: Peace, War, Industry, and the Arts TB/1113
FREDERICK L. NUSSBAUM: The Triumph of Science and Reason, 1660-1685. * Illus. TB/3009
JOHN PLAMENATZ: German Marxism and Russian Communism. º New Preface by the Author TB/1189
RAYMOND W. POSTGATE, Ed.: Revolution from 1789 to 1906: Selected Documents TB/1063
PENFIELD ROBERTS: The Quest for Security, 1715-1740. * Illus. TB/3016
PRISCILLA ROBERTSON: Revolutions of 1848: A Social History TB/1025
ALBERT SOREL: Europe Under the Old Regime. Translated by Francis H. Herrick TB/1121
N. N. SUKHANOV: The Russian Revolution, 1917: Eyewitness Account. Edited by Joel Carmichael Vol. I TB/1066; Vol. II TB/1067
A. J. P. TAYLOR: The Habsburg Monarch, 1809-1918: A History of the Austrian Empire and Austria-Hungary º TB/1187

JOHN B. WOLF: The Emergence of the Great Powers, 1685-1715. * Illus. TB/3010
JOHN B. WOLF: France: 1814-1919: The Rise of a Liberal-Democratic Society TB/3019

Intellectual History & History of Ideas

HERSCHEL BAKER: The Image of Man: A Study of the Idea of Human Dignity in Classical Antiquity, the Middle Ages, and the Renaissance TB/1047
R. R. BOLGAR: The Classical Heritage and Its Beneficiaries: From the Carolingian Age to the End of the Renaissance TB/1125
RANDOLPH S. BOURNE: War and the Intellectuals: Collected Essays, 1915-1919. ‡ Edited by Carl Resek TB/3043
J. BRONOWSKI & BRUCE MAZLISH: The Western Intellectual Tradition: From Leonardo to Hegel TB/3001
ERNST CASSIRER: The Individual and the Cosmos in Renaissance Philosophy. Translated with an Introduction by Mario Domandi TB/1097
NORMAN COHN: The Pursuit of the Millennium: Revolutionary Messianism in medieval and Reformation Europe TB/1037
G. RACHEL LEVY: Religious Conceptions of the Stone Age and Their Influence upon European Thought. Illus. Introduction by Henri Frankfort TB/106
ARTHUR O. LOVEJOY: The Great Chain of Being: A Study of the History of an Idea TB/1009
PERRY MILLER & T. H. JOHNSON, Editors: The Puritans: A Sourcebook of Their Writings
Vol. I TB/1093; Vol. II TB/1094
MILTON C. NAHM: Genius and Creativity: An Essay in the History of Ideas TB/1196
ROBERT PAYNE: Hubris: A Study of Pride. Foreword by Sir Herbert Read TB/1031
RALPH BARTON PERRY: The Thought and Character of William James: Briefer Version TB/1156
BRUNO SNELL: The Discovery of the Mind: The Greek Origins of European Thought TB/1018
PAGET TOYNBEE: Dante Alighieri: His Life and Works. Edited with intro. by Charles S. Singleton TB/1206
ERNEST LEE TUVESON: Millennium and Utopia: A Study in the Background of the Idea of Progress. | New Preface by the Author TB/1134
PAUL VALÉRY: The Outlook for Intelligence TB/2016
PHILIP P. WIENER: Evolution and the Founders of Pragmatism. Foreword by John Dewey TB/1212

Literature, Poetry, The Novel & Criticism

JAMES BAIRD: Ishmael: The Art of Melville in the Contexts of International Primitivism TB/1023
JACQUES BARZUN: The House of Intellect TB/1051
W. J. BATE: From Classic to Romantic: Premises of Taste in Eighteenth Century England TB/1036
RACHEL BESPALOFF: On the Iliad TB/2006
R. P. BLACKMUR et al.: Lectures in Criticism. Introduction by Huntington Cairns TB/2003
ABRAHAM CAHAN: The Rise of David Levinsky: a documentary novel of social mobility in early twentieth century America. Intro. by John Higham TB/1028
ERNST R. CURTIUS: European Literature and the Latin Middle Ages TB/2015
GEORGE ELIOT: Daniel Deronda: a novel. Introduction by F. R. Leavis TB/1039
ETIENNE GILSON: Dante and Philosophy TB/1089
ALFRED HARBAGE: As They Liked It: A Study of Shakespeare's Moral Artistry TB/1035

STANLEY R. HOPPER, Ed.: Spiritual Problems in Contemporary Literature § TB/21
A. R. HUMPHREYS: The Augustan World: Society, Thought and Letters in 18th Century England ° TB/1105
ALDOUS HUXLEY: Antic Hay & The Giaconda Smile. ° Introduction by Martin Green TB/3503
ALDOUS HUXLEY: Brave New World & Brave New World Revisited. ° Introduction by Martin Green TB/3501
HENRY JAMES: Roderick Hudson: a novel. Introduction by Leon Edel TB/1016
HENRY JAMES: The Tragic Muse: a novel. Introduction by Leon Edel TB/1017
ARNOLD KETTLE: An Introduction to the English Novel.
Volume I: Defoe to George Eliot TB/1011
Volume II: Henry James to the Present TB/1012
ROGER SHERMAN LOOMIS: The Development of Arthurian Romance TB/1167
JOHN STUART MILL: On Bentham and Coleridge. Introduction by F. R. Leavis TB/1070
KENNETH B. MURDOCK: Literature and Theology in Colonial New England TB/99
SAMUEL PEPYS: The Diary of Samuel Pepys. ° Edited by O. F. Morshead. Illus. by Ernest Shepard TB/1007
ST.-JOHN PERSE: Seamarks TB/2002
GEORGE SANTAYANA: Interpretations of Poetry and Religion § TB/9
C. P. SNOW: Time of Hope: a novel TB/1040
HEINRICH STRAUMANN: American Literature in the Twentieth Century. Third Edition, Revised TB/1168
PAGET TOYNBEE: Dante Alighieri: His Life and Works. Edited with intro. by Charles S. Singleton TB/1206
DOROTHY VAN GHENT: The English Novel: Form and Function TB/1050
E. B. WHITE: One Man's Meat. Introduction by Walter Blair TB/3505
MORTON DAUWEN ZABEL, Editor: Literary Opinion in America Vol. I TB/3013; Vol. II TB/3014

Myth, Symbol & Folklore

JOSEPH CAMPBELL, Editor: Pagan and Christian Mysteries Illus. TB/2013
MIRCEA ELIADE: Cosmos and History: The Myth of the Eternal Return § TB/2050
C. G. JUNG & C. KERÉNYI: Essays on a Science of Mythology: The Myths of the Divine Child and the Divine Maiden TB/2014
DORA & ERWIN PANOFSKY: Pandora's Box: The Changing Aspects of a Mythical Symbol. Revised Edition. Illus. TB/2021
ERWIN PANOFSKY: Studies in Iconology: Humanistic Themes in the Art of the Renaissance. 180 illustrations TB/1077
JEAN SEZNEC: The Survival of the Pagan Gods: The Mythological Tradition and its Place in Renaissance Humanism and Art. 108 illustrations TB/2004
HELLMUT WILHELM: Change: Eight Lectures on the I Ching TB/2019
HEINRICH ZIMMER: Myths and Symbols in Indian Art and Civilization. 70 illustrations TB/2005

Philosophy

G. E. M. ANSCOMBE: An Introduction to Wittgenstein's Tractatus. Second edition, Revised. ° TB/1210
HENRI BERGSON: Time and Free Will: An Essay on the Immediate Data of Consciousness ° TB/1021

5

H. J. BLACKHAM: Six Existentialist Thinkers: Kierkegaard, Nietzsche, Jaspers, Marcel, Heidegger, Sartre °
TB/1002

CRANE BRINTON: Nietzsche. *New Preface, Bibliography and Epilogue by the Author* TB/1197

ERNST CASSIRER: The Individual and the Cosmos in Renaissance Philosophy. *Translated with an Introduction by Mario Domandi* TB/1097

ERNST CASSIRER: Rousseau, Kant and Goethe. *Introduction by Peter Gay* TB/1092

FREDERICK COPLESTON: Medieval Philosophy ° TB/376

F. M. CORNFORD: Principium Sapientiae: *A Study of the Origins of Greek Philosophical Thought. Edited by W. K. C. Guthrie* TB/1213

F. M. CORNFORD: From Religion to Philosophy: *A Study in the Origins of Western Speculation* § TB/20

WILFRID DESAN: The Tragic Finale: *An Essay on the Philosophy of Jean-Paul Sartre* TB/1030

PAUL FRIEDLÄNDER: Plato: *An Introduction* TB/2017

ÉTIENNE GILSON: Dante and Philosophy TB/1089

WILLIAM CHASE GREENE: Moira: *Fate, Good, and Evil in Greek Thought* TB/1104

W. K. C. GUTHRIE: The Greek Philosophers: *From Thales to Aristotle* ° TB/1008

F. H. HEINEMANN: Existentialism and the Modern Predicament TB/28

ISAAC HUSIK: A History of Medieval Jewish Philosophy
TB/803

EDMUND HUSSERL: Phenomenology and the Crisis of Philosophy. *Translated with an Introduction by Quentin Lauer* TB/1170

IMMANUEL KANT: The Doctrine of Virtue, *being Part II of The Metaphysic of Morals. Trans. with Notes & Intro. by Mary J. Gregor. Foreword by H. J. Paton*
TB/110

IMMANUEL KANT: Groundwork of the Metaphysic of Morals. *Trans. & analyzed by H. J. Paton* TB/1159

IMMANUEL KANT: Lectures on Ethics. § *Introduction by Lewis W. Beck* TB/105

QUENTIN LAUER: Phenomenology: *Its Genesis and Prospect* TB/1169

GABRIEL MARCEL: Being and Having: *An Existential Diary. Intro. by James Collins* TB/310

GEORGE A. MORGAN: What Nietzsche Means TB/1198

PHILO SAADYA GAON, & JEHUDA HALEVI: Three Jewish Philosophers. *Ed. by Hans Lewy, Alexander Altmann, & Isaak Heinemann* TB/813

MICHAEL POLANYI: Personal Knowledge: *Towards a Post-Critical Philosophy* TB/1158

WILLARD VAN ORMAN QUINE: Elementary Logic: *Revised Edition* TB/577

WILLARD VAN ORMAN QUINE: From a Logical Point of View: *Logico-Philosophical Essays* TB/566

BERTRAND RUSSELL et al.: The Philosophy of Bertrand Russell. *Edited by Paul Arthur Schilpp*
Vol. I TB/1095; Vol. II TB/1096

L. S. STEBBING: A Modern Introduction to Logic TB/538

ALFRED NORTH WHITEHEAD: Process and Reality: *An Essay in Cosmology* TB/1033

PHILIP P. WIENER: Evolution and the Founders of Pragmatism. *Foreword by John Dewey* TB/1212

WILHELM WINDELBAND: A History of Philosophy
Vol. I: *Greek, Roman, Medieval* TB/38
Vol. II: *Renaissance, Enlightenment, Modern* TB/39

LUDWIG WITTGENSTEIN: The Blue and Brown Books °
TB/1211

Political Science & Government

JEREMY BENTHAM: The Handbook of Political Fallacies: *Introduction by Crane Brinton* TB/1069

KENNETH E. BOULDING: Conflict and Defense: *A General Theory* TB/3024

CRANE BRINTON: English Political Thought in the Nineteenth Century TB/1071

EDWARD S. CORWIN: American Constitutional History: *Essays edited by Alpheus T. Mason and Gerald Garvey* TB/1136

ROBERT DAHL & CHARLES E. LINDBLOM: Politics, Economics, and Welfare: *Planning and Politico-Economic Systems Resolved into Basic Social Processes* TB/3037

JOHN NEVILLE FIGGIS: The Divine Right of Kings. *Introduction by G. R. Elton* TB/1191

JOHN NEVILLE FIGGIS: Political Thought from Gerson to Grotius: 1414-1625: *Seven Studies. Introduction by Garrett Mattingly* TB/1032

F. L. GANSHOF: Feudalism TB/1058

G. P. GOOCH: English Democratic Ideas in Seventeenth Century TB/1006

J. H. HEXTER: More's Utopia: *The Biography of an Idea. New Epilogue by the Author* TB/1195

ROBERT H. JACKSON: The Supreme Court in the American System of Government TB/1106

DAN N. JACOBS, Ed.: The New Communist Manifesto & Related Documents. *Third edition, Revised* TB/1078

DAN N. JACOBS & HANS BAERWALD, Eds.: Chinese Communism: *Selected Documents* TB/3031

ROBERT GREEN MC CLOSKEY: American Conservatism in the Age of Enterprise, 1865-1910 TB/1137

KINGSLEY MARTIN: French Liberal Thought in the Eighteenth Century: *Political Ideas from Bayle to Condorcet* TB/1114

JOHN STUART MILL: On Bentham and Coleridge. *Introduction by F. R. Leavis* TB/1070

JOHN B. MORRALL: Political Thought in Medieval Times
TB/1076

JOHN PLAMENATZ: German Marxism and Russian Communism. ° *New Preface by the Author* TB/1189

SIR KARL POPPER: The Open Society and Its Enemies
Vol. I: *The Spell of Plato* TB/1101
Vol. II: *The High Tide of Prophecy: Hegel, Marx, and the Aftermath* TB/1102

HENRI DE SAINT-SIMON: Social Organization, The Science of Man, and Other Writings. *Edited and Translated by Felix Markham* TB/1152

JOSEPH A. SCHUMPETER: Capitalism, Socialism and Democracy TB/3008

CHARLES H. SHINN: Mining Camps: *A Study in American Frontier Government.* ‡ *Edited by Rodman W. Paul*
TB/3062

Psychology

ALFRED ADLER: The Individual Psychology of Alfred Adler. *Edited by Heinz L. and Rowena R. Ansbacher*
TB/1154

ALFRED ADLER: Problems of Neurosis. *Introduction by Heinz L. Ansbacher* TB/1145

ANTON T. BOISEN: The Exploration of the Inner World: *A Study of Mental Disorder and Religious Experience*
TB/87

HERBERT FINGARETTE: The Self in Transformation: *Psychoanalysis, Philosophy and the Life of the Spirit.* ‖
TB/1177

SIGMUND FREUD: On Creativity and the Unconscious: *Papers on the Psychology of Art, Literature, Love, Religion.* § *Intro. by Benjamin Nelson* TB/45

C. JUDSON HERRICK: The Evolution of Human Nature TB/545
WILLIAM JAMES: Psychology: *The Briefer Course.* Edited with an Intro. by Gordon Allport TB/1034
C. G. JUNG: Psychological Reflections TB/2001
C. G. JUNG: Symbols of Transformation: *An Analysis of the Prelude to a Case of Schizophrenia.* Illus.
Vol. I: TB/2009; Vol. II TB/2010
C. G. JUNG & C. KERÉNYI: Essays on a Science of Mythology: *The Myths of the Divine Child and the Divine Maiden* TB/2014
JOHN T. MC NEILL: A History of the Cure of Souls TB/126
KARL MENNINGER: Theory of Psychoanalytic Technique TB/1144
ERICH NEUMANN: Amor and Psyche: *The Psychic Development of the Feminine* TB/2012
ERICH NEUMANN: The Archetypal World of Henry Moore. 107 illus. TB/2020
ERICH NEUMANN: The Origins and History of Consciousness Vol. I Illus. TB/2007; Vol. II TB/2008
C. P. OBERNDORF: A History of Psychoanalysis in America TB/1147
RALPH BARTON PERRY: The Thought and Character of William James: *Briefer Version* TB/1156
JEAN PIAGET, BÄRBEL INHELDER, & ALINA SZEMINSKA: The Child's Conception of Geometry ° TB/1146
JOHN H. SCHAAR: Escape from Authority: *The Perspectives of Erich Fromm* TB/1155

Sociology

JACQUES BARZUN: Race: *A Study in Superstition.* Revised Edition TB/1172
BERNARD BERELSON, Ed.: The Behavioral Sciences Today TB/1127
ABRAHAM CAHAN: The Rise of David Levinsky: *A documentary novel of social mobility in early twentieth century America.* Intro. by John Higham TB/1028
THOMAS C. COCHRAN: The Inner Revolution: *Essays on the Social Sciences in History* TB/1140
ALLISON DAVIS & JOHN DOLLARD: Children of Bondage: *The Personality Development of Negro Youth in the Urban South* || TB/3049
ST. CLAIR DRAKE & HORACE R. CAYTON: Black Metropolis: *A Study of Negro Life in a Northern City.* Revised and Enlarged. Intro. by Everett C. Hughes
Vol. I TB/1086; Vol. II TB/1087
EMILE DURKHEIM et al.: Essays on Sociology and Philosophy: *With Analyses of Durkheim's Life and Work.* || Edited by Kurt H. Wolff TB/1151
LEON FESTINGER, HENRY W. RIECKEN & STANLEY SCHACHTER: When Prophecy Fails: *A Social and Psychological Account of a Modern Group that Predicted the Destruction of the World* || TB/1132
ALVIN W. GOULDNER: Wildcat Strike: *A Study in Worker-Management Relationships* || TB/1176
FRANCIS J. GRUND: Aristocracy in America: *Social Class in the Formative Years of the New Nation* TB/1001
KURT LEWIN: Field Theory in Social Science: *Selected Theoretical Papers.* || Edited with a Foreword by Dorwin Cartwright TB/1135
R. M. MACIVER: Social Causation TB/1153
ROBERT K. MERTON, LEONARD BROOM, LEONARD S. COTTRELL, JR., Editors: Sociology Today: *Problems and Prospects* || Vol. I TB/1173; Vol. II TB/1174
TALCOTT PARSONS & EDWARD A. SHILS, Editors: Toward a General Theory of Action: *Theoretical Foundations for the Social Sciences* TB/1083

JOHN H. ROHRER & MUNRO S. EDMONSON, Eds.: The Eighth Generation Grows Up: *Cultures and Personalities of New Orleans Negroes* || TB/3050
ARNOLD ROSE: The Negro in America: *The Condensed Version of Gunnar Myrdal's An American Dilemma* TB/3048
KURT SAMUELSSON: Religion and Economic Action: *A Critique of Max Weber's The Protestant Ethic and the Spirit of Capitalism.* || ° *Trans.* by E. G. French; Ed. with Intro. by D. C. Coleman TB/1131
PITIRIM A. SOROKIN: Contemporary Sociological Theories. *Through the First Quarter of the 20th Century* TB/3046
MAURICE R. STEIN: The Eclipse of Community: *An Interpretation of American Studies* TB/1128
FERDINAND TÖNNIES: Community and Society: *Gemeinschaft und Gesellschaft.* Translated and edited by Charles P. Loomis TB/1116
W. LLOYD WARNER & Associates: Democracy in Jonesville: *A Study in Quality and Inequality* TB/1129
W. LLOYD WARNER: Social Class in America: *The Evaluation of Status* TB/1013

RELIGION

Ancient & Classical

J. H. BREASTED: Development of Religion and Thought in Ancient Egypt. Introduction by John A. Wilson TB/57
HENRI FRANKFORT: Ancient Egyptian Religion: *An Interpretation* TB/77
G. RACHEL LEVY: Religious Conceptions of the Stone Age and their Influence upon European Thought. Illus. Introduction by Henri Frankfort TB/106
MARTIN P. NILSSON: Greek Folk Religion. Foreword by Arthur Darby Nock TB/78
ALEXANDRE PIANKOFF: The Shrines of Tut-Ankh-Amon. Edited by N. Rambova. 117 illus. TB/2011
H. J. ROSE: Religion in Greece and Rome TB/55

Biblical Thought & Literature

W. F. ALBRIGHT: The Biblical Period from Abraham to Ezra TB/102
C. K. BARRETT, Ed.: The New Testament Background: *Selected Documents* TB/86
C. H. DODD: The Authority of the Bible TB/43
M. S. ENSLIN: Christian Beginnings TB/5
M. S. ENSLIN: The Literature of the Christian Movement TB/6
JOHN GRAY: Archaeology and the Old Testament World. Illus. TB/127
H. H. ROWLEY: The Growth of the Old Testament TB/107
D. WINTON THOMAS, Ed.: Documents from Old Testament Times TB/85

The Judaic Tradition

LEO BAECK: Judaism and Christianity. *Trans.* with Intro. by Walter Kaufmann TB/823
SALO W. BARON: Modern Nationalism and Religion TB/818
MARTIN BUBER: Eclipse of God: *Studies in the Relation Between Religion and Philosophy* TB/12
MARTIN BUBER: Moses: *The Revelation and the Covenant* TB/27
MARTIN BUBER: Pointing the Way. Introduction by Maurice S. Friedman TB/103
MARTIN BUBER: The Prophetic Faith TB/73
MARTIN BUBER: Two Types of Faith: *the interpenetration of Judaism and Christianity* ° TB/75

ERNST LUDWIG EHRLICH: A Concise History of Israel: From the Earliest Times to the Destruction of the Temple in A.D. 70 ° TB/128
MAURICE S. FRIEDMAN: Martin Buber: The Life of Dialogue TB/64
LOUIS GINZBERG: Students, Scholars and Saints TB/802
SOLOMON GRAYZEL: A History of the Contemporary Jews TB/816
WILL HERBERG: Judaism and Modern Man TB/810
ABRAHAM J. HESCHEL: God in Search of Man: A Philosophy of Judaism TB/807
ISAAC HUSIK: A History of Medieval Jewish Philosophy TB/803
FLAVIUS JOSEPHUS: The Great Roman-Jewish War, with The Life of Josephus. Introduction by William R. Farmer TB/74
JACOB R. MARCUS The Jew in the Medieval World TB/814
MAX L. MARGOLIS & ALEXANDER MARX: A History of the Jewish People TB/806
T. J. MEEK: Hebrew Origins TB/69
C. G. MONTEFIORE & H. LOEWE, Eds.: A Rabbinic Anthology TB/832
JAMES PARKES: The Conflict of the Church and the Synagogue: The Jews and Early Christianity TB/821
PHILO, SAADYA GAON, & JEHUDA HALEVI: Three Jewish Philosophers. Ed. by Hans Lewey, Alexander Altmann, & Isaak Heinemann TB/813
HERMAN L. STRACK: Introduction to the Talmud and Midrash TB/808
JOSHUA TRACHTENBERG: The Devil and the Jews: The Medieval Conception of the Jew and its Relation to Modern Anti-Semitism TB/822

Christianity: General

ROLAND H. BAINTON: Christendom: A Short History of Christianity and its Impact on Western Civilization. Illus. Vol. I, TB/131; Vol. II, TB/132

Christianity: Origins & Early Development

AUGUSTINE: An Augustine Synthesis. Edited by Erich Przywara TB/335
ADOLF DEISSMANN: Paul: A Study in Social and Religious History TB/15
EDWARD GIBBON: The Triumph of Christendom in the Roman Empire (Chaps. XV-XX of "Decline and Fall," J. B. Bury edition). § Illus. TB/46
MAURICE GOGUEL: Jesus and the Origins of Christianity.° Introduction by C. Leslie Mitton
 Volume I: Prolegomena to the Life of Jesus TB/65
 Volume II: The Life of Jesus TB/66
EDGAR J. GOODSPEED: A Life of Jesus TB/1
ADOLF HARNACK: The Mission and Expansion of Christianity in the First Three Centuries. Introduction by Jaroslav Pelikan TB/92
R. K. HARRISON: The Dead Sea Scrolls: An Introduction ° TB/84
EDWIN HATCH: The Influence of Greek Ideas on Christianity. § Introduction and Bibliography by Frederick C. Grant TB/18
ARTHUR DARBY NOCK: Early Gentile Christianity and Its Hellenistic Background TB/111
ARTHUR DARBY NOCK: St. Paul ° TB/104
JAMES PARKES: The Conflict of the Church and the Synagogue: The Jews and Early Christianity TB/821
SULPICIUS SEVERUS et al.: The Western Fathers: Being the Lives of Martin of Tours, Ambrose, Augustine of Hippo, Honoratus of Arles and Germanus of Auxerre. Edited and translated by F. R. Hoare TB/309

F. VAN DER MEER: Augustine the Bishop: Church and Society at the Dawn of the Middle Ages TB/304
JOHANNES WEISS: Earliest Christianity: A History of the Period A.D. 30-150. Introduction and Bibliography by Frederick C. Grant Volume I TB/53
 Volume II TB/54

Christianity: The Middle Ages and The Reformation

JOHANNES ECKHART: Meister Eckhart: A Modern Translation by R. B. Blakney TB/8
DESIDERIUS ERASMUS: Christian Humanism and the Reformation: Selected Writings. Edited and translated by John C. Olin TB/1166
ÉTIENNE GILSON: Dante and Philosophy TB/1089
WILLIAM HALLER: The Rise of Puritanism TB/22
JOHAN HUIZINGA: Erasmus and the Age of Reformation. Illus. TB/19
A. C. MCGIFFERT: Protestant Thought Before Kant. Preface by Jaroslav Pelikan TB/93
JOHN T. MCNEILL: Makers of the Christian Tradition: From Alfred the Great to Schleiermacher TB/121
G. MOLLAT: The Popes at Avignon, 1305-1378 TB/308
GORDON RUPP: Luther's Progress to the Diet of Worms ° TB/120

Christianity: The Protestant Tradition

KARL BARTH: Church Dogmatics: A Selection TB/95
KARL BARTH: Dogmatics in Outline TB/56
KARL BARTH: The Word of God and the Word of Man TB/13
RUDOLF BULTMANN et al.: Translating Theology into the Modern Age: Historical, Systematic and Pastoral Reflections on Theology and the Church in the Contemporary Situation. Volume 2 of Journal for Theology and the Church, edited by Robert W. Funk in association with Gerhard Ebeling TB/252
WINTHROP HUDSON: The Great Tradition of the American Churches TB/98
SOREN KIERKEGAARD: Edifying Discourses. Edited with an Introduction by Paul Holmer TB/32
SOREN KIERKEGAARD: The Journals of Kierkegaard. ° Edited with an Introduction by Alexander Dru TB/52
SOREN KIERKEGAARD: The Point of View for My Work as an Author: A Report to History. § Preface by Benjamin Nelson TB/88
SOREN KIERKEGAARD: The Present Age. § Translated and edited by Alexander Dru. Introduction by Walter Kaufmann TB/94
SOREN KIERKEGAARD: Purity of Heart TB/4
SOREN KIERKEGAARD: Repetition: An Essay in Experimental Psychology. Translated with Introduction & Notes by Walter Lowrie TB/117
SOREN KIERKEGAARD: Works of Love: Some Christian Reflections in the Form of Discourses TB/122
WALTER LOWRIE: Kierkegaard: A Life Vol. I TB/89
 Vol. II TB/90
PERRY MILLER & T. H. JOHNSON, Editors: The Puritans: A Sourcebook of Their Writings Vol. I TB/1093
 Vol. II TB/1094
JAMES M. ROBINSON et al.: The Bultmann School of Biblical Interpretation: New Directions? Volume 1 of Journal of Theology and the Church, edited by Robert W. Funk in association with Gerhard Ebeling TB/251
F. SCHLEIERMACHER: The Christian Faith. Introduction by Richard R. Niebuhr Vol. I TB/108
 Vol. II TB/109

F. SCHLEIERMACHER: On Religion: *Speeches to Its Cultured Despisers.* Intro. by Rudolf Otto   TB/36
PAUL TILLICH: Dynamics of Faith   TB/42
EVELYN UNDERHILL: Worship   TB/10
G. VAN DER LEEUW: Religion ... Essence and Manifestation: *A Study in Phenomenology.* Appendices by Hans H. Penner   Vol. I TB/100; Vol. II TB/101

Christianity: The Roman and Eastern Traditions

A. ROBERT CAPONIGRI, Ed.: Modern Catholic Thinkers I: *God and Man*   TB/306
A. ROBERT CAPONIGRI, Ed.: Modern Catholic Thinkers II: *The Church and the Political Order*   TB/307
THOMAS CORBISHLEY, S. J.: Roman Catholicism   TB/112
CHRISTOPHER DAWSON: The Historic Reality of Christian Culture   TB/305
G. P. FEDOTOV: The Russian Religious Mind: *Kievan Christianity, the 10th to the 13th centuries*   TB/70
G. P. FEDOTOV, Ed.: A Treasury of Russian Spirituality   TB/303
DAVID KNOWLES: The English Mystical Tradition   TB/302
GABRIEL MARCEL: Being and Having: *An Existential Diary.* Introduction by James Collins   TB/310
GABRIEL MARCEL: Homo Viator: *Introduction to a Metaphysic of Hope*   TB/397
GUSTAVE WEIGEL, S. J.: Catholic Theology in Dialogue   TB/301

Oriental Religions: Far Eastern, Near Eastern

TOR ANDRAE: Mohammed: *The Man and His Faith*   TB/62
EDWARD CONZE: Buddhism: *Its Essence and Development.* ° Foreword by Arthur Waley   TB/58
EDWARD CONZE et al., Editors: Buddhist Texts Through the Ages   TB/113
ANANDA COOMARASWAMY: Buddha and the Gospel of Buddhism. *Illus.*   TB/119
H. G. CREEL: Confucius and the Chinese Way   TB/63
FRANKLIN EDGERTON, Trans. & Ed.: The Bhagavad Gita   TB/115
SWAMI NIKHILANANDA, Trans. & Ed.: The Upanishads: *A One-Volume Abridgment*   TB/114
HELLMUT WILHELM: Change: *Eight Lectures on the I Ching*   TB/2019

Philosophy of Religion

NICOLAS BERDYAEV: The Beginning and the End §   TB/14
NICOLAS BERDYAEV: Christian Existentialism: *A Berdyaev Synthesis.* Ed. by Donald A. Lowrie   TB/130
NICOLAS BERDYAEV: The Destiny of Man   TB/61
RUDOLF BULTMANN: History and Eschatology: *The Presence of Eternity* °   TB/91
RUDOLF BULTMANN AND FIVE CRITICS: Kerygma and Myth: *A Theological Debate*   TB/80
RUDOLF BULTMANN and KARL KUNDSIN: Form Criticism: *Two Essays on New Testament Research.* Translated by Frederick C. Grant   TB/96
MIRCEA ELIADE: The Sacred and the Profane   TB/81
LUDWIG FEUERBACH: The Essence of Christianity. § Introduction by Karl Barth. Foreword by H. Richard Niebuhr   TB/11
ADOLF HARNACK: What is Christianity? § Introduction by Rudolf Bultmann   TB/17

FRIEDRICH HEGEL: On Christianity: *Early Theological Writings.* Ed. by R. Kroner & T. M. Knox   TB/79
KARL HEIM: Christian Faith and Natural Science   TB/16
IMMANUEL KANT: Religion Within the Limits of Reason Alone. § Intro. by T. M. Greene & J. Silber   TB/67
JOHN MACQUARRIE: An Existentialist Theology: *A Comparison of Heidegger and Bultmann.* ° Preface by Rudolf Bultmann   TB/125
PAUL RAMSEY, Ed.: Faith and Ethics: *The Theology of H. Richard Niebuhr*   TB/129
PIERRE TEILHARD DE CHARDIN: The Phenomenon of Man °   TB/83

Religion, Culture & Society

JOSEPH L. BLAU, Ed.: Cornerstones of Religious Freedom in America: *Selected Basic Documents, Court Decisions and Public Statements. Revised and Enlarged Edition*   TB/118
C. C. GILLISPIE: Genesis and Geology: *The Decades before Darwin* §   TB/51
KYLE HASELDEN: The Racial Problem in Christian Perspective   TB/116
WALTER KAUFMANN, Ed.: Religion from Tolstoy to Camus: *Basic Writings on Religious Truth and Morals. Enlarged Edition*   TB/123
JOHN T. MCNEILL: A History of the Cure of Souls   TB/126
KENNETH B. MURDOCK: Literature and Theology in Colonial New England   TB/99
H. RICHARD NIEBUHR: Christ and Culture   TB/3
H. RICHARD NIEBUHR: The Kingdom of God in America   TB/49
RALPH BARTON PERRY: Puritanism and Democracy   TB/1138
PAUL PFUETZE: Self, Society, Existence: *Human Nature and Dialogue in the Thought of George Herbert Mead and Martin Buber*   TB/1059
WALTER RAUSCHENBUSCH: Christianity and the Social Crisis. ‡ Edited by Robert D. Cross   TB/3059
KURT SAMUELSSON: Religion and Economic Action: *A Critique of Max Weber's* The Protestant Ethic and the Spirit of Capitalism. || ° Trans. by E. G. French; Ed. with Intro. by D. C. Coleman   TB/1131
ERNST TROELTSCH: The Social Teaching of the Christian Churches °   Vol. I TB/71; Vol. II TB/72

NATURAL SCIENCES AND MATHEMATICS

Biological Sciences

CHARLOTTE AUERBACH: The Science of Genetics Σ   TB/568
MARSTON BATES: The Natural History of Mosquitoes. *Illus.*   TB/578
A. BELLAIRS: Reptiles: *Life History, Evolution, and Structure. Illus.*   TB/520
LUDWIG VON BERTALANFFY: Modern Theories of Development: *An Introduction to Theoretical Biology*   TB/554
LUDWIG VON BERTALANFFY: Problems of Life: *An Evaluation of Modern Biological and Scientific Thought*   TB/521
HAROLD F. BLUM: Time's Arrow and Evolution   TB/555
JOHN TYLER BONNER: The Ideas of Biology. Σ *Illus.*   TB/570
A. J. CAIN: Animal Species and their Evolution. *Illus.*   TB/519
WALTER B. CANNON: Bodily Changes in Pain, Hunger, Fear and Rage. *Illus.*   TB/562

W. E. LE GROS CLARK: The Antecedents of Man: *Intro. to Evolution of the Primates.* ° *Illus.* TB/559
W. H. DOWDESWELL: Animal Ecology. *Illus.* TB/543
W. H. DOWDESWELL: The Mechanism of Evolution. *Illus.* TB/527
R. W. GERARD: Unresting Cells. *Illus.* TB/541
DAVID LACK: Darwin's Finches. *Illus.* TB/544
J. E. MORTON: Molluscs: *An Introduction to their Form and Functions. Illus.* TB/529
ADOLF PORTMANN: Animals as Social Beings. ° *Illus.* TB/572
O. W. RICHARDS: The Social Insects. *Illus.* TB/542
P. M. SHEPPARD: Natural Selection and Heredity. *Illus.* TB/528
EDMUND W. SINNOTT: Cell and Psyche: *The Biology of Purpose* TB/546
C. H. WADDINGTON: How Animals Develop. *Illus.* TB/553

Chemistry

J. R. PARTINGTON: A Short History of Chemistry. *Illus.* TB/522
J. READ: A Direct Entry to Organic Chemistry. *Illus.* TB/523
J. READ: Through Alchemy to Chemistry. *Illus.* TB/561

Communication Theory

J. R. PIERCE: Symbols, Signals and Noise: *The Nature and Process of Communication* TB/574

Geography

R. E. COKER: This Great and Wide Sea: *An Introduction to Oceanography and Marine Biology. Illus.* TB/551
F. K. HARE: The Restless Atmosphere TB/560

History of Science

W. DAMPIER, Ed.: Readings in the Literature of Science. *Illus.* TB/512
A. HUNTER DUPREE: Science in the Federal Government: *A History of Policies and Activities to 1940* TB/573
ALEXANDRE KOYRÉ: From the Closed World to the Infinite Universe: *Copernicus, Kepler, Galileo, Newton, etc.* TB/31
A. G. VAN MELSEN: From Atomos to Atom: *A History of the Concept Atom* TB/517
O. NEUGEBAUER: The Exact Sciences in Antiquity TB/552
H. T. PLEDGE: Science Since 1500: *A Short History of Mathematics, Physics, Chemistry and Biology. Illus.* TB/506
HANS THIRRING: Energy for Man: *From Windmills to Nuclear Power* TB/556
LANCELOT LAW WHYTE: Essay on Atomism: *From Democritus to 1960* TB/565
A. WOLF: A History of Science, Technology and Philosophy in the 16th and 17th Centuries. ° *Illus.*
Vol. I TB/508; Vol. II TB/509
A. WOLF: A History of Science, Technology, and Philosophy in the Eighteenth Century. ° *Illus.*
Vol. I TB/539; Vol. II TB/540

Mathematics

H. DAVENPORT: The Higher Arithmetic: *An Introduction to the Theory of Numbers* TB/526
H. G. FORDER: Geometry: *An Introduction* TB/548

GOTTLOB FREGE: The Foundations of Arithmetic: *A Logico-Mathematical Enquiry* TB/534
S. KÖRNER: The Philosophy of Mathematics: *An Introduction* TB/547
D. E. LITTLEWOOD: Skeleton Key of Mathematics: *A Simple Account of Complex Algebraic Problems* TB/525
GEORGE E. OWEN: Fundamentals of Scientific Mathematics TB/569
WILLARD VAN ORMAN QUINE: Mathematical Logic TB/558
O. G. SUTTON: Mathematics in Action. ° *Foreword by James R. Newman. Illus.* TB/518
FREDERICK WAISMANN: Introduction to Mathematical Thinking. *Foreword by Karl Menger* TB/511

Philosophy of Science

R. B. BRAITHWAITE: Scientific Explanation TB/515
J. BRONOWSKI: Science and Human Values. *Revised and Enlarged Edition* TB/505
ALBERT EINSTEIN et al.: Albert Einstein: Philosopher-Scientist. *Edited by Paul A. Schilpp* Vol. I TB/502
Vol. II TB/503
WERNER HEISENBERG: Physics and Philosophy: *The Revolution in Modern Science* TB/549
JOHN MAYNARD KEYNES: A Treatise on Probability. ° *Introduction by N. R. Hanson* TB/557
SIR KARL POPPER: The Logic of Scientific Discovery TB/576
STEPHEN TOULMIN: Foresight and Understanding: *An Enquiry into the Aims of Science. Foreword by Jacques Barzun* TB/564
STEPHEN TOULMIN: The Philosophy of Science: *An Introduction* TB/513
G. J. WHITROW: The Natural Philosophy of Time ° TB/563

Physics and Cosmology

STEPHEN TOULMIN & JUNE GOODFIELD: The Fabric of the Heavens: *The Development of Astronomy and Dynamics. Illus.* TB/579
DAVID BOHM: Causality and Chance in Modern Physics. *Foreword by Louis de Broglie* TB/536
P. W. BRIDGMAN: The Nature of Thermodynamics TB/537
P. W. BRIDGMAN: A Sophisticate's Primer of Relativity TB/575
A. C. CROMBIE, Ed.: Turning Point in Physics TB/535
C. V. DURELL: Readable Relativity. *Foreword by Freeman J. Dyson* TB/530
ARTHUR EDDINGTON: Space, Time and Gravitation: *An outline of the General Relativity Theory* TB/510
GEORGE GAMOW: Biography of Physics Σ TB/567
MAX JAMMER: Concepts of Force: *A Study in the Foundation of Dynamics* TB/550
MAX JAMMER: Concepts of Mass *in Classical and Modern Physics* TB/571
MAX JAMMER: Concepts of Space: *The History of Theories of Space in Physics. Foreword by Albert Einstein* TB/533
EDMUND WHITTAKER: History of the Theories of Aether and Electricity
Volume I: *The Classical Theories* TB/531
Volume II: *The Modern Theories* TB/532
G. J. WHITROW: The Structure and Evolution of the Universe: *An Introduction to Cosmology. Illus.* TB/504